ROUTLEDGE LIBRARY EDITIONS
HUMAN G

T0227490

Volume 7

INSTITUTIONS AND
GEOGRAPHICAL PATTERNS

INSTITUTIONS AND GEOGRAPHICAL PATTERNS

Edited by
ROBIN FLOWERDEW

Routledge
Taylor & Francis Group

LONDON AND NEW YORK

First published in 1982 by Croom Helm Ltd

This edition first published in 2016
by Routledge
2 Park Square, Milton Park, Abingdon, Oxon OX14 4RN

and by Routledge
711 Third Avenue, New York, NY 10017

Routledge is an imprint of the Taylor & Francis Group, an informa business

British Library Cataloguing in Publication Data
A catalogue record for this book is available from the British Library

ISBN: 978-1-138-95340-6 (Set)
ISBN: 978-1-315-65887-2 (Set) (ebk)
ISBN: 978-1-138-95508-0 (Volume 7) (hbk)
ISBN: 978-1-315-66659-4 (Volume 7) (ebk)

Publisher's Note
The publisher has gone to great lengths to ensure the quality of this reprint but
points out that some imperfections in the original copies may be apparent.

Disclaimer
The publisher has made every effort to trace copyright holders and would welcome
correspondence from those they have been unable to trace.

Institutions and Geographical Patterns

EDITED BY ROBIN FLOWERDEW

CROOM HELM
London & Canberra

British Library Cataloguing in Publication Data

Institutions and geographical patterns.—(Croom
    Helm series in geography and environment)
    1. Anthropo-geography—Congresses
    I. Flowerdew, Robin
    304.2   GF41

    ISBN 0-7099-1011-8

Printed and bound in Great Britain by
Biddles Ltd, Guildford and King's Lynn

CONTENTS

LIST OF FIGURES

LIST OF TABLES

FOREWORD

This book grew from a recognition that an increasing
amount of work in many aspects of human geography
was concerned with the effects of institutions of
different types . The theme of institutional involvement seemed to take rather different forms in
different parts of geography - interesting work in
the study of housing was concerned with the operation of building societies and local authorities;
in agricultural geography there was concern about
the activity of pension funds and other financial
institutions in the land market; in other areas
there was increasing interest in the effects of
local government policies on local welfare, and
the effects of planning mechanisms on the outcomes
of controversial decisions.
     Likewise, my experience in explaining aspects
of the geography of the USA to British students
led me to the realisation that institutional
differences between the two countries were of considerable importance in explaining social and
economic differences.   The differences between
British and American urban problems were directly
related to differences in the powers and responsibilities of local government; American environmental
policy differed from British because of the
differences in the role of the courts in interpreting law and the degree of Civil Service discretion.
     The 1980 meeting of the Institute of British
Geographers was held at Lancaster, and my colleagues
and I were asked to consider what theme, if any,
the Conference should have.   For the reasons just
stated, my suggested theme was the role of institutions in shaping geographical patterns.   In the
event, it was decided not to adopt a single
Conference theme, but Richard Auty, the local
secretary, encouraged me to convene a session on the
topic, and this duly took place in January 1980.
     One of the contributors to this session
suggested that publication of the papers, and others
on the same theme, would be worthwhile, and I discussed this possibility with Michael Bradford, who
is Advisory Editor of Croom Helm's Geography series.
It is largely due to his interest and encouragement

that we went ahead with the idea of publication.
Some of the contributors to the Lancaster session
were unable to contribute to the book, but others
were recruited, and the scope of the book is considerably greater than that of the conference
session.

The book is intended to be disciplinary - all
the authors have trained as geographers - but it
is intended to bring together work from many
different subdisciplines of geography. A good
deal of ink has been used in stressing the importance of interdisciplinary contact - quite
rightly - and the bibliographies of all the papers
in this book include many items from outside
geography, but perhaps there is a need too for
stronger intradisciplinary contact within human
geography. It is probably easier today to
persuade an industrial geographer that new ideas in
industrial economics are important to him or her
than that developments in political geography are
important. Ideas which are well-known, or even
passé, in one branch of geography may be just as
relevant in another branch, where it is virtually
unknown. An example relevant to this book might
be the managerialist approach, well-known in urban
sociology and urban geography, but seldom discussed
in, say, rural geography or the geography of
tourism. The essays in the book are intended to
be connected by the theme of institutions, and it
is hoped that readers from each subdiscipline of
human geography may learn not only from the chapter
concerned with that subdiscipline, but from the
ideas and concepts discussed in other chapters.

The contributors selected were people I knew
to be doing interesting work with an institutional
theme.   Some attempt was made to spread the contributions out between different parts of geography,
but it will be seen that urban geography is well-
represented, and some topics, like transport
geography and third-world development, have been
omitted.   It seems as if institutionally-related
work is much more common in some parts of geography
than others, with urban geography leading the way.

Each author has adopted a somewhat different
approach to the study of institutional effects.
Some are concerned solely with government and the
effects of government policy, while others discuss
business and financial institutions.   They vary
in other aspects of their approach, some working
within a marxist framework, some in an empiricist
tradition, and others attempting to synthesise

insights from several different viewpoints. Some
would see a focus on institutions as an important
development in geographical methodology, and others
see them merely as a factor that must be considered
to achieve an adequate understanding of reality.

I would like to thank the contributors for
producing what I believe are interesting and worthwhile essays on an important topic, and for making
the editorial task so problem-free. My colleagues
at Lancaster and at Rutgers, where the final version
of the book was put together, helped with their
ideas and encouragement, especially Richard Auty
and Tom Manion. Michael Bradford's interest was
probably critical in getting the venture off the
ground, and Peter Sowden of Croom Helm was helpful
in overseeing the progress of the book.

I am very grateful to Anne Jackson who
prepared the illustrations. My thanks also to
Dawn Phazey at Lancaster and to Jackie O'Leary and
Loretta Bishop at Rutgers for their work on intermediate stages of manuscript preparation.
Marian Jackson was invaluable in proofreading and
preparing the index. My greatest debt, however,
is to Christine Skinner, who typed the manuscript
in camera-ready form, working cheerfully and
skilfully under severe time constraints. I have
been lucky in having such pleasant and capable
people to work with.

Chapter 1

INTRODUCTION : INSTITUTIONAL APPROACHES IN GEOGRAPHY

Tom Manion and Robin Flowerdew

In this volume, the study of institutions is advocated as an approach to the explanation of geographical patterns. Many geographers, along with members of other disciplines, have studied the effects of institutions on various aspects of society, the economy and the environment, and this approach has led to significant insights in several parts of the subject. Some parts of the subject, such as urban geography, have been familiar with an institutional approach under the name 'urban managerialism' for several years; in other parts of the field, work has proceeded along similar lines without being recognised as a separate approach; but in other parts insufficient attention has been given to the institutional theme. Advocating the study of institutions, then,is not especially new in itself.

What is perhaps more original, however, is to present the study of institutions as a theme of relevance to much or all of human geography. An advantage of such a presentation may be to encourage links between work exploring institutional themes in different substantive contexts, so that, for example, geographers interested in the role of institutions in urban geography may share and profit from the work of those studying institutional effects on environmental protection or land-use patterns. In recent years much stress has been placed on the benefits of inter-disciplinary contacts; perhaps there is also a need for greater contact between the different parts of the discipline of geography.

An institutional approach seeks to explain phenomena of geographical interest through stress on one type of factor, the effect of institutional structure and actions. As such, it is an alternative or supplement to attempts at explanation based on individual preference and choice, on the workings of the free market system, or on conflict between social classes. It is not a philosophy, and should be compatible with several philosophical perspectives, including 'the scientific method', dialectical materialism, and at least some brands of humanism. Again, it is not a methodology, and is compatible with econometric modelling, systems analysis, survey sampling, detailed case-study or participant observation. Later in this chapter, and later in the book, some methodological strategies and problems will be reviewed and exemplified.

This introduction is divided into five main sections. First, the concept of an 'institutional approach' will be elaborated, with a discussion of other kinds of approach with which it may be compatible. Although we think that most of the contributors to the book would agree with most of this statement, we should stress that this is a personal view, and that some or all of them may disagree with our view of what constitutes an institutional approach, and when and if it is a valid approach to adopt.

Secondly, we will trace the development of institutionally based work in geography. Although the discussion is intended to be relevant to all fields of human geography, the emphasis is often on urban geography and housing. This is because we feel that methodological discussion in urban geography (and related areas like urban sociology) has had more relevance to institutional studies than discussion in other parts of the discipline, and that many of the issues discussed in urban geography, such as the role of managers in the urban system and the theory of the local state, could fruitfully be considered in other parts of geography. A fuller account of some of these issues in the study of housing is available from Bassett and Short (1980).

A third topic is the relation of institutional approaches in geography to various developments in the other social sciences, including institutional economics, post-Keynesian economics, and political economy. In these schools of thought, attention has been directed away from the simplifying assumptions frequently made by other approaches in economics, political science or sociology towards a more realistic appraisal of how institutions mediate, transform or replace the processes traditionally regarded as central.

The fourth goal is to pick out a series of themes relevant to the study of institutional activities and

their geographical repercussions. These will include theories and generalisations which have been
applied to institutional behaviour, types of research
which have been carried out, and problems which
confront the student of institutional activity.

Lastly, the chapters of the book will be discussed briefly, with the intention of highlighting
similarities and differences in approach. Emphasis
will be given to bringing out ideas or findings that
are relevant to more than one subdivision of
geography.

The Institutional Approach

Geography since the quantitative revolution of the
1960s has been troubled by conflicts between two
types of explanatory goal: the search for generality
or theory on one hand and the desire for relevance
to specific applications on the other. Much dissatisfaction with influential theoretical approaches
has been caused by their inability to incorporate
the realities which the layman can identify as
important in specific places. A retreat from theory
to a string of unconnected case studies seems just
as unsatisfactory. The need is for general statements which shed light on specific cases, and for
theory based on assumptions which are not simplified
to the point of unreality.

A major reason for the difficulty in generalis-
ing from specific cases is the variation in institutional constraints from place to place and from time
to time. A local economy, for example, operates in
a complex institutional framework, created and
regulated partly by government, central and local,
and partly by totally or partially autonomous organisations, such as banks and building societies,
which control or influence access to capital and the
types of goods and services produced. Similar
comments can be made in urban, social, environmental
and other aspects of geography. Attempts to create
general theory have fallen foul of the institutional
and organisational constraints operating in the real
world. Such factors are not regarded as a source
of error in applying a theory, or as a signal to
return to idiographic ad hoc explanations. Rather,
an institutional approach to human geography seeks
to focus attention on these factors and to develop
theory concerning the impact of institutional controls on the activities of individuals or companies,
and on the spatial processes and patterns generated

3

by these activities.

The term 'institution' refers to an organisation considered in relation to the effects of its internal structure and operating constraints on how it acts. The organisation may be a private company, a nationalised industry, a non-profit organisation, a central or local government or one of its departments, or a quango (quasi-autonomous non-governmental organisation). It is not regarded as an indivisible entity acting as a unit to maximise its own utility or to fulfil its mission. It is regarded as being influenced by rules and procedures which are not directly and essentially related to its mission, but which have been developed internally or imposed externally on it. It will also be influenced by its own internal structure and by the special interests of its subdivisions or of people within it.

An institutional approach, then, seeks for explanation of events through a study of the institutions affecting them, with stress on the effects of the rules, procedures, and internal structure of these institutions. It does not deny that the actions of many institutions are directed towards their expressed goals; however, it stresses that the ways in which they are organised and operate may affect their actions and impacts. It also recognises that actions may have important indirect, perhaps unintended, consequences as well as those for which they were designed. It does not deny that individual people make choices and that these choices affect their own welfare and that of others; it stresses that choices are made under constraints, and that often the organisation and policies of institutions may restrict the choices available. It does not deny that social processes reflect the access of classes to the means of production; it asserts that this access is mediated through specific institutional arrangements, and that any realistic understanding of our society must depend on the study of institutions if it is to achieve significant detail.

The study of institutions can be relevant to geography in several ways. First the goals of institutions can be important in their effects on operations, whether the goals are explicit or unstated. In particular, an institution whose ostensible purpose is to facilitate transactions between individuals may, by its mere existence, acquire its own interests in growth or self-perpetuation. It may also be important to study the individuals

involved in implementing the institution's policies.
The values and professional background of these
people may affect how these policies are put into
practice.   Secondly, institutions may set up a
system of regulations, and the form and details of
these may have major effects on spatial outcomes.
The choice of one method of regulation rather than
another may have a considerable influence on the
activities concerned, and on the ways in which
people attempt to sidestep the regulations.   Thirdly,
some institutions may operate in ways that have
wholly unintended consequences, through a failure
to reconcile conflicting objectives or through unexpected responses from the people concerned.   In
particular, individuals or groups may be able to
manipulate laws or sets of regulations so as to
achieve results very different from the intentions
of those setting up the institutional structure.
Fourthly, organisations may create their own
internal structures - for example, the local and
regional units employed by most large companies and
by governments - and the nature of the regional
division may reflect decentralisation of policy-
making and implementation, and may guide the allocation of resources.   Organisational as well as
spatial structure may have important effects on
geographical patterns and distributions.
      Examples of geographical work taking an insti-
tutional approach might therefore include studies of
the spatial impact of a particular piece of legislation, a court decision, or a governmental bureau
or agency; the allocation of government expenditure
between regions; the effects of different local
government policies and procedures on their areas
of jurisdiction; the effects of the administrative
and spatial organisation of public facilities and of
large companies; the effects of the policies of
banks and lending and investment institutions on
the economic patterns of the areas in which they
operate; and the actions of trade unions, employers'
organisations, or special interest groups and their
impact on economic or land-use policy.
      The field in which an institutional approach
has been most clearly and explicitly developed is
urban geography.   Here the sociologist R.E. Pahl
has developed the approach of 'urban managerialism',
where emphasis is placed on the operation of housing
market institutions and their managers in regulating
people's access to housing in different areas and of
different types.   A considerable literature has
developed discussing urban managerialism and its

strengths and weaknesses in comparison to other approaches to the study of urban housing. The term 'institutional' is preferred here to 'managerial', as the latter seems to direct too much attention to the role of the managers themselves as opposed to the constraints within which they have to work.

A distinction should also be made here between geographical work done for institutions - solicited or unsolicited advice - and work about institutions. Although the geographical study of institutions is often relevant, perhaps essential, to policy questions, its aim is not to help in policy forma- tion, but to describe, analyse, and construct theories concerning the impact of policies. Using a distinction from economics, the approach is positive, not normative - though not necessarily positivistic in its philosophical presuppositions.

The Development of Institutional Approaches in Geography

Geographers have until recently been reluctant to consider the existence of institutional effects in their attempts to explain empirical reality. This was essentially true for traditional regional geography, where descriptions of countries tended to say very little about the political and administra- tive structure there. Regional textbooks of the 1960s continued to stress traditional societies and the colonially-based export economy at a time when independence and the associated transformation of institutions were creating new economic and social patterns in most of the Third World. Similarly, accounts of economic geography traditionally dis- cussed types of manufacturing industry and trade without even naming the companies involved, let alone considering the effects of their corporate structure.

This lack of specificity and disregard for political factors may be linked to the old theme of geography as the study of how the physical environ- ment has affected man. In this approach, the economic and cultural geography of a region were regarded as being closely related to its resources, and hence largely independent of political circum- stances. An extreme expression of this is Sauer's statement (1918, pp. 421-2): 'Geology is indeed a subject of every day importance in Tennessee, for upon it depend the wealth or poverty, progress or stagnation, even the political convictions of its sections to an extraordinary degree.' Cultural geography studies tended to be concentrated in rural and remote areas, where the importance of environmental influences on man seemed less open to question, and where institutional effects could more easily be attributed to a superorganic entity called 'culture', adapted to the environment. Even industrial location could be explained by the existence of certain factors like raw material availability, market, or labour supply. As such it was not very relevant which company was located at which place; the locational factors would automatically attract that industry to that place, and the source of the capital and the relationship of the plant in question to the entire corporate strategy was not regarded as important.

Early Interest in Institutions

A few exceptions can be found to these generalisations. One of the earliest geographers to consider institutional factors was Whittlesey, in his paper on 'The Impress of Effective Central Authority upon the Landscape' (1935). He drew attention to the effects of political boundary changes (exemplified by the growth of Gdynia, Poland, in the inter-war period when Danzig became politically separate from the Vistula valley); the tendency of governments to impose uniformity on the whole of their territories, through land survey or through building style in government offices; the ability of governments to stimulate poorer areas by redirecting funds from richer areas; and the effects of tariffs and regulations on economic geography.

Nelson (1952) was one of the first geographers to study the effects of municipal boundaries in urban areas. He raised the question (p. 178): 'are local municipal boundaries reflected in the urban land use pattern? ... If so, the local political complex within urban areas is an essential part of the geographic understanding of modern cities.' He concluded from a study of greater Los Angeles that these boundaries were important. Large (1957) and Fielding (1964; 1965) discussed the influence of government activities on agriculture, while Manners (1962) provided a relatively early discussion of government effects on industrial location, pointing out, among other things, the lack of cohesion between British regional policy, aimed at decentralisation, and the motorway programme, which encouraged centralisation.

The Quantitative Revolution

Studies of this type, however, were relatively unusual until the later 1960s, and were overshadowed
by the methodological and philosophical changes
occurring at the same time.   In the earlier part
of the quantitative revolution, economic factors,
rather than institutional, were stressed.   Of
particular importance were studies aiming to relate
economic process to spatial form, with central place
theory and location theory of especial significance.
The economic background to geographical models of
this type came from neoclassical economics, and so
these types of model tended to adopt simplifying
assumptions similar to those made in neoclassical
economics - notably those of perfect competition and
individual rationality.   Under perfect competition,
there is a large number of small firms competing
without restriction.   In models of this type, the
institutional context supposes the existence of a
market in which clearance occurs, but no other institutions, governments, stock exchanges, or anything
else, affect the operation of the firms or of the
market.   Each firm is considered atomistically, and
so their internal organisation is also disregarded.
As a result of these influences, models in
economic geography tended to put little emphasis on
institutional effects.   As is sometimes the case
in neoclassical economics, institutions were regarded
as 'nuisance factors', means of explaining the
failure of models to explain the detail of observed
reality.   As such, their treatment was extra-
theoretical and little attempt was made to model
their effects.   A partial exception may have been
the study of socialist countries, where it was
obvious that government policies did not fit in with
the largely institution-free assumptions of neoclassical economics.   Writers on the geography of
the USSR and other socialist countries had to face
the problem of trying to understand the effects of
the institutional structures set up there; however,
little attempt seems to have been made at constructing general theory about the relationships between
institutions and geographical reality.
In urban geography, researchers attempted to
match the achievements of economic geography by
reasoning from economic process to spatial form,
especially through the development of economic

models of urban spatial structure. Others tried to derive spatial form from the study of social processes, especially in urban ecology, derived from the work of Burgess and Park. Burgess' neo-Darwinian theory of groups in competition for residential space also had little to say about the effects of institutions on urban structure, although the Chicago school of urban ecology has had a good deal to say about the effects of community institutions on maintaining the cohesiveness of social groups, especially European immigrant groups. The processes of invasion and succession, however, required little institutional context to make them work beyond the existence of a housing market where social groups could compete on the basis of ability to pay and willingness to accept overcrowded conditions.

The success of urban ecology models in explaining urban structure in the United States in the early twentieth century can perhaps be related to the comparative freedom from government control (as well as from the effects of existing urban development) in that time and that nation. Institutional factors could again be suggested as reasons for its failure in other contexts - such as the existence of publicly-owned housing or planning and zoning laws. In fact, it was an institutional change, in the regulations governing immigration to the USA, that eventually caused a reduction in the flow of immigrants that provided the dynamic for Burgess' model of competition for living space.

Behavioural Approaches

An important criticism of urban ecology and economic models in the later 1960s was that they were unrealistic. Models based on social process suggested that social groups behaved in a particular way, without making it clear how this was reflected in individual behaviour. Economic models did make assumptions about individual behaviour, but these assumptions were based on a clearly unrealistic picture of man, who was regarded as having full knowledge, perfect computational ability and a goal of utility-maximisation (directly translated in economic geography as profit-maximisation).

Behavioural geography developed in reaction to this lack of realism. It was perhaps most concerned with examining the implications of the difference between the real decision-making environment and

individuals' perceptions of it.   In addition, the
concept of satisficing behaviour   (developed by
Simon, 1959, in the context of organisational study)
was adopted as an alternative to utility maximisa-
tion, and decision-making was studied in terms of
the relative weighting given by people to the
attributes of those alternatives they were choosing
among.

Despite the link with early work on organisational behaviour, most behavioural geography concentrated on individual people and had little to
say about the effects of institutions on spatial
behaviour.   A partial exception is the literature
on industrial location decision-making, where
Townroe (1971) examined the structure of the
decision process in some detail.   He discussed the
structure of the decision in relation to the amount
of resources employed, and to the location of the
decision-maker within the corporate hierarchy.

In another paper (1972), Townroe was able to
derive some fairly general hypotheses about organisational behaviour.   These included hypotheses
about the effect of company structure, for example,
that 'privately owned concerns are more open to
subjective pressures than publicly owned companies'
(p. 270), and about the behaviour of decision-makers,
for example, that 'decisions undertaken and arrived
at by single individuals involve less formal commitment to specific goals and less formal appraisal of
alternatives' and that 'dependence upon external
finance does not result in externally imposed conditions' (p. 270).   Presaging a theme of importance
to the whole of this book, he felt that concepts
derived from organisation theory were of limited
usefulness because they are concerned with routin-
ised procedures: 'there is little satisfactory
theoretical grounding (except perhaps what can be
borrowed from psychology) for the study of individuals acting within an organizational context, but
in a situation of stress without an accustomed
frame of reference' (p. 271).   The more complex
organisational structure of multi-plant enterprises
has been discussed by Watts (1974).   For such
organisations, internal structure may have major
effects on locational inertia or change (see also
Lloyd and Dicken, 1977, pp. 374-83).

Choices and Constraints

The stress on individual preferences, decision-

making and action in behavioural geography led to a
concern with institutions in another way.   Most
geographers interested in the behavioural approach
were not concerned with individual behaviour per se,
but about the aggregate patterns to which individual
actions contributed.   From their viewpoint, individual preferences and choices did not seem to provide
sufficient insight into the aggregate reality.   This
was because behavioural models left out of account
the question of what determined the set of alternatives among which individuals were choosing.   In
the case of residential mobility, it was obvious
that most people did not live in their ideal, most-
preferred home.   Part of the reason was expense,
and part was the existence of institutional constraints (availability of a mortgage, qualification
for publicly supplied housing, perhaps racial discrimination).   In addition, the supply of housing
was limited, and itself was governed by the institutional and other constraints on the housing industry.
    Some of these criticisms of the behavioural
approach were expressed by Gray (1975, pp. 228,230),
contrasting an institutional approach to a
behavioural one:

> The dominant explanatory referent underlying
> work by urban geographers is the notion that
> people exercise individual preferences made
> within a choice framework.   Thus geographers
> seek to explain residential structures, the
> type and location of housing occupied by people
> and patterns of household mobility on the basis
> that individuals and families 'choose' where
> and in what type of housing they live ...
> Instead many groups are constricted and con-
> strained from choice and pushed into particular
> housing situations because of their position in
> the housing market and by individuals and
> institutions controlling the operation of
> particular housing systems.

Rather than assuming that residential differentiation represents differences in choice and preference
associated with differences in culturally determined
behaviour and values, Gray points out that the
housing system does not give everybody free choice.
He feels it is more appropriate to examine differentiation from a position which recognises the spatial
distribution of housing opportunities as an imposed
pattern often conflicting with preference.   Access
to housing and to resources generally is seen as

11

determined by status and power in relation to
allocative institutions whose operation generates
constraints which in turn preclude choice maximisa-
tion.   This type of approach directly challenges
many of the assumptions underlying the behavioural
tradition by stressing societal institutions and
their active role in shaping urban areas.

The Relevance Debate

Another important development in geography in the
late 1960s and early 1970s was the increased interest
in 'relevance'.   Often this was associated with
liberal or radical political views, and a desire to
apply geographical research to improving the condition
of mankind.   For others, it grew out of the wish
that geography could be recognised as relevant to
policy-making and policy evaluation by government.
In both cases, geographers wanting to be relevant had
to consider what forms of institutional (usually
governmental) action could alter the outcome of a
process.
     Coppock (1974) argued for a greater role for
geographers in public policy, stressing not only the
ignorance of policy-makers about the potential contribution of geographers,but also the lack of awareness by geographers that 'there is virtually no
aspect of contemporary geography which is not affected
to some degree by public policy.' (p. 5)   Thus
awareness of the processes of policy formation was
not only worthwhile in itself but important in
giving geographers a complete picture of factors
affecting their ordinary fields of interest.
     As Johnston (1979, chapter 6) shows, interest
in 'relevance' was interpreted in many different
ways by different geographers, from those who argued
that the removal of injustices and inequalities required both revolutionary theory and revolutionary
practice (Folke, 1972) to those who discussed the
advisability of charging for countryside recreation
(McCallum and Adams, 1980).   In some cases, the
topic was a relatively minor change in institutional
policy or procedure, in others a fundamental change
in the nature of the institutions affecting people's
lives.
     Policy-relevant study in geography can be
classified according to how critically institutional
activity is treated.   It can range from studies of
various strategies that an institution, or government agency, can employ to meet its stated goals,

through studies of the effects of changes in the
structure or composition of the institution or of
the laws or other institutions governing its
operation, to studies of the impact of totally new
or transformed institutions on the society or
economy that they affect.

Approaches can also vary as to how far they
accept the official view of the aims and functions of
policies. Some interesting studies have revealed
how institutional policies may have effects
different from those stated (Gray, 1976, shows how
local authority criteria of tenant suitability are
used to reinforce social stratification by allocating lower-status tenants to the worst estates),
effects different from those intended, or effects
produced by interaction with apparently unrelated
policies (Robson, 1975, shows how British taxation
and housing policies have worked together to reduce
the private rented sector in the housing market).
Other studies have shown some of the effects when
policies have been put into effect partially or not
at all (Davies, 1972, shows how 'planning blight'
can arise in this way).

This increased interest in government policy
was often prescriptive in intention, but good
research in the field had to include theoretical and
empirical arguments to justify policy evaluations
and recommendations. Much work on the effects of
government policy has been generated by motives of
this kind.

Urban Managerialism

The study of institutions in geography can also be
linked to the work of urban sociologists, especially
Form (1954), Rex and Moore (1967) and Pahl (1970).
Their work was done in reaction to urban ecology and
neoclassical urban economics. Form (1954) rejected
the idea that urban land use was a response to un-
fettered market forces: 'the image of a free and
unorganised market in which individuals compete
impersonally for land must be abandoned. The
reason for this is that the land market is highly
organised and dominated by a number of interacting
organisations.' He viewed land use as the outcome
of conflict and bargaining between these organisations, and suggested that the activities of these
organisations depended on their resources, their
internal organisation and the interests to which
they are accountable.

The beginnings of the managerial approach to urban housing can be traced more directly to Rex and Moore's study, Race, Community and Conflict (1967). Rex and Moore used concepts from Weberian sociology in developing their critique of urban ecology. Ecological studies suggested that the forces 'sifting and sorting' (Park, 1952, p. 16) individuals throughout the city were those of the market. The nature of the market, however, was not sufficiently analysed; rather it was accepted uncritically. Park, for example, alludes to the forces of competition and conflict arising from the market system in the following way (1952, p. 16): 'The metropolis is it seems a great sifting and sorting mechanism which in ways not yet fully understood infallibly selects out of the population as a whole the individuals best suited to live in a particular region or milieu.'

Rex and Moore (1967) focused their analysis directly on the 'sifting and sorting' mechanisms, viewing the city as a social system wherein scarce resources are distributed between classes and consumed by them. Rex and Moore's main thesis was that there is a 'class struggle over the use of houses and that this class struggle is the central process of the city as a social unit' (Rex and Moore, 1967, p. 273). Given a relatively fixed supply of desirable residences and a continuous demand for them there will be competition and conflict between individuals in their attempts to secure access to housing. They also suggested that this process may occur independently of the individual's labour market position because often individuals in the same labour market position will have differential degrees of access to housing. Defining six housing classes Rex and Moore suggested that certain groups and individuals, such as West Indians, are compelled to occupy particular housing because of their weak position in relation to the administrators of housing opportunities.

This explicitly Weberian framework aroused much debate and has been widely criticised (Davies and Taylor, 1970; Haddon, 1970; Pahl, 1975; Lambert, Paris and Blackaby, 1978). Haddon (1970) suggested that Rex and Moore confused categories of the population with types of housing and paid too little attention to the means of access. In other words, Rex and Moore's categorisation of housing classes is the end product of the workings of the housing market and does not reflect the ability of individuals to negotiate the rules of eligibility. As Haddon (1970, p. 129) notes, Rex approaches this

formulation somewhat tentatively.   'In this process
people are distinguished from one another by their
strength in the housing market or, more generally,
in the system of housing allocation' (Rex, 1968,
p. 214).   Although Rex and Moore (1967) concentrate
on the definition of housing classes, nevertheless
the introduction of conflict, constraints, the
notion of housing class and the emphasis upon the
social distribution of housing opportunities represented a more critical, deeper analysis than the
urban ecology approach which preceded it.
      Pahl (1970, pp. 215-22) claimed first that
there are fundamental spatial constraints on access
to scarce urban resources and facilities.
Secondly, he claimed that there are fundamental
social constraints on access to scarce urban
resources and facilities which reflect the distribution of power in the society.   These constraints
mould the socio-spatial, or socio-ecological, system,
within which managers and 'gatekeepers' collectively,
through the operation of bureaucratic rules and
procedures, distribute and control the consumption
of urban resources thereby affecting the access of
the population to them.   Pahl termed the former
(managers and urban gatekeepers) the independent
variable, and the latter (access) the dependent
variable.   Pahl's work is directly in line with
that of Haddon (1970, p. 132) who suggested that
people will have differential access to housing,
depending on the power derived from their market
situation, status or political influence.   Pahl
reaffirmed Haddon's (1970, p. 133) stress on the
need for analysis of the means of access and the
ways that different people can negotiate the institutions and constraints of the housing market.   As
previously noted the notion of constraints on
housing opportunities meant that studies of residential differentiation and spatial structure changed
their emphasis, stressing eligibility and supply
factors more than choice, aspiration or preference.
Access to housing was seen as determined by wealth,
status or power in relation to institutions allocating housing resources.   Essentially Pahl (1970)
suggested that a better understanding of urban
phenomena would be gained through a focus upon the
institutions and individuals who control the distribution of societal resources.
      There have been many studies that have used
Pahl's framework directly and indirectly.   Lambert
(1970) for example referred to the managers of the
urban system as the independent variable and the

position of immigrants as the dependent variable.
He analysed the development of racial segregation
in Birmingham during the 1960s from what may be
termed an early managerialist position.   Other
studies of this nature include Davies (1972) on
planners in Newcastle-upon-Tyne; Dennis (1970) on
the sociology of housing in Sunderland; Hatch (1973)
on the ideological and organisational backgrounds of
building society managers; Paris (1974) on public
housing officials and planners in Birmingham;
Corina (1976) on allocation policy and practice in
Oldham; English (1976) on allocation to deprived
estates on Clydeside; Gray (1976) on selection and
allocation in the public sector in Hull; Williams
(1976) on the role of building societies and estate
agents in Islington, Niner (1975) on local authority
policy and practice in the West Midlands; and
Randall (1981) on the contribution of housing
officers to policy-making in London boroughs.
These studies represent important empirical developments within the context of Pahl's original formulation.   Although some of them are not explicitly
managerialist, the approach taken is based on the
recognition that the controllers of access exert an
important influence upon the distribution of
housing opportunities.
     Outside the realm of housing and urban studies,
other studies were conducted to investigate the
people who controlled the allocation of resources.
In environmental management, Sewell (1971) investigated the training and values of people involved in
making decisions about environmental quality.   He
found that engineers tended to perceive water
pollution as an economic problem requiring a technological answer, whereas public health officials
perceived it as a health problem requiring the
imposition of regulations.   Both groups were sceptical of the potential contribution of the public or
of other professional groups to the definition
and solution of problems.   Sewell related these
attitudes to sociological work concerning the
effects on an individual's view about the problems
he or she deals with of his or her identification
with the organisation he or she works for (Gerstl
and Hutton, 1966).   Mitchell (1971) also observed
differences between a group of professionals in
water resource management and the general public in
their views of water pollution and the role of the
public in the managerial process.   The questionnaire
approach to managerial perceptions has also been
applied to manufacturing industry by McDermott and

Taylor (1976), who collected evaluations from 528
managers in New Zealand of the importance of
sixteen attributes of local business environments;
they were able to relate the responses to firm
characteristics.

The managerialist approach has been criticised
on the grounds that the relationship of managers and
managed is greatly constrained by broader economic
and social processes (Norman, 1975). This point
can be illustrated by the work of Harloe,
Issacharoff and Minns (1974). Their study of
housing in two London boroughs, Sutton and Lambeth,
provides a description of the inter-relationships
between the public and private sector under a plura-
listic model of resource control and allocation.
In this model derived from American political
scientists and historians such as Dahl (1961; 1963),
Lipset (1969) and Hofstadter (1966), decision-
making is taken to be the outcome of the interplay
between competing group interests and views within
a community. The authors emphasise that their
work is concerned both with ideology and constraints
and attempt to 'move one step back in this study ...
while we outline the ideologies and goals of organisations and suggest how these can act as constraints
affecting people, we also concentrate on what the
constraints facing organisations tell us about the
constraints facing people' (Harloe et al., 1974,
p. 9). They illustrate some of the constraints
(high land prices, inter- and intra-authority
conflicts) preventing Sutton and Lambeth from
making rapid short-term improvements in housing
conditions. As Norman (1975) notes, this study
clarifies the determinants of power in the housing
market and shows how the interaction of ideology and
constraints produces differentiated housing
opportunities.

Similarly the work of Byrne and Davis (1974)
represents an attempt at understanding the interplay
between competing groups. They suggest that
nationally and locally resource allocation is
primarily intended to serve the interests of private
capital, and that decisions taken by planners and
other groups are explicable in terms of the class
interests they foster, interests which may be dia-
metrically opposed to those of the local community.
'A good deal of what the Tynemouth Borough Council
proposed for the future of this area could probably
be explained in terms of the consequences of a
hypothetical member of staff of Poulson and
Associates getting out of his car on a sunny

afternoon, looking out across Shields harbour, and saying "this view is too good for the working-classes."' (1974, p. 22)

Both studies represent attempts to get away from the restricted focus of earlier managerialism on urban managers themselves to analyses of power relations in the context of the broader socio-political and economic structure. Lambert, Blackaby and Paris (1975) also recognise the limitations of what they call 'naive managerialism' referring to the ideological and political aspects of management.

Pahl's original discussion (1970) provoked much debate in sociology and geography (Harloe, 1975; Williams, 1978; Leonard, 1979). One implication of this debate was that a focus upon managers per se implicitly assigns them independence and autonomy without recognising the wider societal structure within which they operate, or the influence that other groups have on how managers can distribute resources. Accepting this critique Pahl reconsidered his earlier position: 'whilst a focus on these urban managers, or gatekeepers, is a useful research strategy and whilst an exploration of their implicit goals, values, assumptions and ideologies may provide a valuable approach for students exploring the role of professionals in bureaucracies, such an approach lacks both practical policy implications and theoretical substance.' (1975, p. 265) He argued that the managerial focus is impractical because its logical conclusion is likely to be that more resources and greater sensitivity in allocation are needed if existing socio-spatial differentials are to be reduced. The contention is that as resources are limited, and are likely to remain so, then managers and institutions are severely constrained and are consequently allowed few alternatives but to operationalise inequitable policies in a utilitarian manner.

In his 1975 work, Pahl had become sceptical of the idea that managerial behaviour is an independent variable. Studies relating findings principally to institutional policies have to make the dubious assumption that managers are essentially independent of outside constraints. This implies, for example, that the political complexion of local councils and other groups involved in the formulation of procedures for the allocation of resources is not influential in the process as a whole. Similarly problems arise when bureaucratic selection, allocation and control are abstracted from the environment

of decision-making - the relationships between urban
managers and the state in capitalist society.

However Pahl suggests that, as urbanism entails
the geographical concentration of a socially designated surplus product and that as urban managers
mediate between urban populations and the capitalist
economy, their role remains crucial to an understanding of urban society.   He concludes that as
cities exist so too will the inequitable generation
of indirect wages through the allocation of
resources by managers.   'The urban managers remain
the allocators of this surplus; they must remain
central to the urban problematic.' (Pahl, 1975,
p. 285)   More recently Pahl has said that irrespective of the theoretical position one adopts regarding class in a capitalist society 'independent
analysis of the agents of allocation is still necessary.'   Further he notes, 'Sadly I fear that so
long as the surplus has to be concentrated, in order
to be redistributed, so long will rules have to be
generated and agents have to apply them.   Some may
think the managerialist thesis is dead; whether or
not that is so I am quite certain that the managers
will not lie down.' Pahl (1979a, p. 89)

Pahl has acknowledged the criticisms of urban
managerialism and has extended his earlier ideas
accordingly.   However despite modification and ela-
boration to include the relationships between urban
managers and government his commitment to a managerial focus remains.   Managerialism has been extended
from a deterministic cause and effect approach by
incorporation of 'higher-order' phenomena such as
the economic, political and ideological structures
of the society.   Thus Pahl has advocated, in a statement not dissimilar from Lasswell's (1958) definition of politics, that 'those interested in this
field should never be far from the practical questions of who gets what, who determines who gets what
and what determines who determines who gets what.
Monocausal answers will be increasingly unlikely.'
(1977a, p. 16)

This broadening of scope should be viewed in
conjunction with theoretical developments concerned
with understanding the role of the state in capitalist society and attempts to relate urban phenomena
to the central economic and political institutions of
contemporary society.   As mentioned previously, the
work of Harloe et al. (1974) and Byrne and Davis
(1974) moved away from early managerialism by arguing that managers were not an 'independent variable'
but acted within constraints set by wider societal

influences.   The concern with power relations and
the constraints restricting the actions of institutions provides a sufficient justification for this
widening of approach.   The institutional approach
being put forward here incorporates both the
managerialist insights into the role of gatekeepers
in implementing the allocation of resources and the
analysis of the critics of managerialism in its
stress on the institutional structures, including the
central and local state, as well as major commercial
and political organisations, that constrain the
'middle-dogs'.   The managers have discretion within
sets of guidelines and constraints; the relative
power they possess can be gauged by empirical study,
and not by a priori assumption.

Institutional Approaches in Cultural Geography

Study of institutions has been advocated in a very
different context by Wagner (1975).   As a cultural
geographer, Wagner was concerned to find a suitable
unit for analysis of human culture, more widespread
and contemporary than the small isolated communities
traditionally studied in cultural geography and
less vague than a tribe or a people.   Institutions
represent distinctive and organised social forms,
usually serving an explicit human purpose and
operating according to definite rules.   Most
human behaviour derives its significance from its
institutional context.   Institutions are responsible for transmitting culture, and institutional
forms constitute a large part of the culture to be
transmitted.   Wagner is particularly concerned with
the relation of man and environment arguing that
'the role of human beings operating institutionally
to change or maintain their environments becomes
the primary concern of cultural ecology' (p. 13).
He laments the lack of geographical studies 'of
individual business concerns, recreational
associations, military units, churches, or families
as elements of cultural geography' (p. 13).
     Cultural geography has been particularly prone
to the preoccupation with environmental influences
on human behaviour discussed at the start of this
chapter.   It has also been governed by a holistic
conception of a superorganic entity called 'culture',
which is the object of study in cultural geography
and which is not reducible to the individuals who
make up the cultural groups (a point of view
examined and demolished by Duncan, 1980).

Environmental influences should not be seen as affecting this superorganic concept, nor a collection of isolated and independent individuals, but a highly organised set of people, whose institutional structures, whether political, legal, economic or recreational, mediated the man-land relationship. Different reactions by different cultures to a similar environment may well be related to the ability of their respective institutions to co-ordinate activities or to supply motivation sufficient to meet the environmental challenge.

The relevance of institutional behaviour to cultural norms and to human behaviour generally is a vast field, and the theme is applicable to many phenomena of interest to geographers. One good example is tourism, where relatively little work in geography or elsewhere has directed attention to 'the social organisation of the support institutions of hotels, airlines, and tour companies' (Schudson, 1979, p. 1257). As Schudson goes on to point out, 'they are not only support institutions, but institutions which invent new attractions to support themselves'.

In other contexts, too, institutions are highly influential in affecting how people organise their time, and institutional policies and activities are thus relevant to studies of time budgets, activity patterns, and other topics normally considered as behavioural geography. Institutional policies about opening hours, of social security offices, hospital outpatients' departments, and ordinary banks or shops may be as effective as distance and accessibility in determining whether an individual has access to facilities and in constraining his or her activities. Discretionary travel is also likely to be controlled by institutional influences, either through the decisions reached by organisations to which individuals belong, or through the supply of possible destinations made available by public or private institutions.

Theories of the State

From much of the preceding discussion, it should be clear that one type of institution, the state, has a particularly important role in its effect on geographical patterns. This section is therefore devoted to a discussion of the nature of central and local government and its influence on the spatial and social distribution of resources. Again, the study

of housing is taken as the main empirical example.
Several recent approaches to understanding the
nature of the state have come from the field of
political economy within which distinctive schools
of thought are identifiable.   Holloway and
Picciotto (1978) discuss the ways in which political
theorists and economists have analysed the state and
society in a review of the current debate in Marxist
political economy.   It is evident from the discussion that distinctions need to be drawn between
theorists working in this area, principally
'politicists' and 'reductionists'.   Politicists
such as Poulantzas (1969) and Miliband (1969) insist
on the relative autonomy of the state and treat it
as a discrete aspect of capitalist society.   This is
in part a reaction to reductionism and the 'over-
simplification of the relation between the economic
and political which presents the political as a mere
reflection of the economic' (Holloway and Picciotto,
1978, p. 6).

Reductionist or economic determinist interpretations make a distinction between the economic and
the political analysing the former to explain the
latter.   Such interpretations are further divided
between neo-Ricardians and fundamentalists (Holloway
and Picciotto, 1978).   Neo-Ricardians accept the
distinction between economic and political and
analyse the nature of the state in terms of its
activities and functions in the economic sphere
(Glyn and Sutcliffe, 1972; Gough, 1975).   Conversely
fundamentalists deny the autonomy of the political,
focusing instead upon the processes of capital
accumulation to explain state activities.

Dear and Clark (1978) summarise a number of
theories of the state and enumerate five alternative
roles that the state may take on.   Although it is
convenient analytically to distinguish between these
roles they should not be regarded as mutually exclusive.   The state can be considered (p. 174) as
a:

1. supplier of public or social goods;
2. regulator and facilitator of the market
    place;
3. social engineer, in the sense of interven-
    ing in the economy to achieve its own
    policy objectives;
4. arbiter between competing social groups or
    classes; and
5. agent within society and the economy on
    behalf of some ruling elite.

The particular form and function that the state has, whether it has any autonomy, and if so how much, will continue to be an important debate in political economy (Holloway and Picciotto, 1978) and in geography (Dear and Clark, 1978; Harvey, 1976), where the journal Environment and Planning A has recently devoted a special issue to the theme (Dear, 1981). The role of the state as arbiter has been investigated with respect to the effects of the legal system on geographical patterns (Teitz, 1978; Clark, 1981), a theme which could well be developed in other contexts.

The continued growth of the public sector in many national economies means that the state is having an increasing impact upon the social and economic well-being of the populace. The state controls considerable resources which are distributed differentially to populations defined socially and spatially (Bennett, 1980). In the United Kingdom for example central and local government expenditure amounts to two-fifths of the national income (Klein, Barnes, Buxton and Craven, 1974) and, as Jackson (1976) points out, local authorities are responsible for the manipulation of funds well in excess of the national defence budget. Further, Prest (1978) shows that in terms of capital expenditure local authorities play a more significant role than central government because of the wide range of goods and services they provide.

The role of institutions mediating between the state and local populations is also increasing. Institutional perspectives are therefore of increasing pertinence, in studying who gets what, where, when and why, and in examining the influence of institutions on the socio-spatial structure of urban areas. The links between private institutions and the local and national state must be investigated in order to understand their respective roles in resource provision. These relationships are essential to linking observable geographical patterns to the social, economic and political processes generating them.

The Local State

The degree of independence from central authority possesed by local governments has also been the subject of recent debate (Dear and Clark, 1981). One interpretation suggests that the 'local state' (Cockburn, 1977) is dominated by the national and

governed by centralised decisions. Green (1959)
suggested 'from the standpoint of local government,
this process (the advent of the welfare state and
the growth of central planning) has not only deprived the minor local authorities of their erstwhile
significance, but, in the interests of tidy administration has changed the character of the major
authorities to that of virtual agents of the central
departments.' A similar point was made by Robson
(1966): 'They, the local authorities, are no longer
partners in a common enterprise but only junior
partners or mere administrative agents of the
centre.' Cockburn (1977), referring to Redcliffe-
Maud (1974), added: 'The misapprehension lies in
the belief that local councils spring from some
ancient right of grass roots self-government.
This is not the case. They are, and under capitalism have always been, subject to central
government.' (p. 46)
    In contrast to the above assertions is a view
which suggests that any discussion of the extent
and nature of local autonomy must be specific to
the aspect of local authority behaviour under
examination. Boaden and Alford made this point
with respect to housing and education suggesting
that 'the timing and scope of local government
services is far more within the scope of their powers
than is commonly recognised in the literature on
English Local Government' (1969, p. 204).
Similarly Dear and Clark (1978) point out the dis-
tinctiveness of local authorities. This is
particularly relevant in the context of municipal
housing, which varies in amount, in its spatial
distribution, and in accessibility for various
social groups. Municipal housing in Britain is a
good example of a policy area where wide discretion
has been permitted in the determination of needs
and policies with central government involved to a
limited extent only. Clearly local government
institutions are not autonomous agents because they
cannot be isolated from the wider social, political
and economic context in which they operate nor from
the operation of financial, commercial, industrial
and property capital (Lamarche, 1976). Indeed as
Broadbent (1977) has noted the links between local
government and the local urban economy are much less
direct than between the national state and the
national economy. Nevertheless, local authorities
in Britain are able to control the allocation of
municipal housing to the social groups in their
jurisdiction, which affects the socio-spatial urban

structure.   Statutory obligations have been imposed
on local governments by legislation, but their
interpretation was considered to be the prerogative
of local authorities as early as the Addison Act,
1919, when the principle that they should have discretion in access and allocation was established
(McKay and Cox, 1979).

This discretion can be seen in the differences
between local authorities in their policies for
allocation of municipal housing.   The variation in
eligibility criteria: housing points, date of
application, and their local interpretation makes
it impossible to find two authorities that are the
same in every respect.   This reflects the national
policy of encouraging local discretion.   The
reluctance to introduce centralised statutory guidelines was reaffirmed in a recent Green Paper: 'The
government wish to maintain their policy of giving
the maximum freedom to authorities to interpret and
implement national policies in the light of local
circumstances.   They therefore believe that proposals to control local authorities' allocations
policies centrally, for example, by laying down a
statutory framework, should be rejected.'
(Department of the Environment, 1977, p. 79)

The management of the municipal housing sector
operates with minimum legal control allowing local
discretion and autonomy.   Centralised guidelines
have taken the form of advisory, rather than super-
visory, circulars and reports.   Central government
has relied on persuasion which in part stems from
the view that local authorities best know the local
context,how to meet needs and fulfil management responsibilities.   As Grant (1976, p. 1) notes,
'Local authorities have almost a free hand in
devising policies to suit their own areas,particular
housing problems, resources, political beliefs,
aspirations and prejudices.'

It is difficult to accept Cockburn's dismissal
of local autonomy in light of this example.   When
discussing the role of the state in society it is
important to specify what aspect of the state is
being discussed.   Rather than approaching the state
as a monolith it seems more appropriate to acknow-
ledge that vertical and horizontal specialisation
has occurred because of the scale and complexity of
bureaucratic organisation.   The state as allocator,
financier, organiser, arbiter and legislator has
concentrated power in order to achieve and manage
societal goals (Pickvance, 1978).   Concomitantly
this has necessitated divisions horizontally into

government departments and vertically into central and local government, the latter of which is itself subdivided into specialist departments. Such divisions have been accompanied by the delegation of specific functions attached to which are certain responsibilities, financial, legislative and ideological.

Pahl (1975, pp. 270-1) has outlined a typology of alternative models of resource control and allocation. While these activities do not cover the entire spectrum of local government, they do provide 'perhaps the most observable indicators of the working of particular authorities' (Kirby, 1979, p. 8). Each of his models produces different explanations of the socio-political process of resource allocation and provides a suitable theoretical framework to which empirical studies may be related (although it should be stressed that the models are not mutually exclusive). Many studies can be related to Pahl's typology. For example Harloe et al. (1974) relates to the pluralist model, Byrne and Davis (1974) the control by capitalist model; Cockburn (1977) the statist model and Gray (1976) the managerialist model. Pahl's typology provides a useful tool for dealing with variations in local authorities' behaviour in a sector where needs are ubiquitous and minimum standards exist but discretion is possible.

Inevitably individuals located in different parts of a country will have differential access to publicly supplied goods because of variability in supply and the bureaucratic rules, eligibility criteria and general procedures operating in different areas. Housing opportunities will vary temporally, spatially and socially in a heterogeneous sector and between a population differentiated in terms of need, determined loosely by statute and more specifically by local authorities - need being a function of the individual's wealth, status and power in relation to the socio-economic structure. Pahl summarises this point (1979b, pp. 43-4):

> Whether the position I adopt is termed managerialist, corporatist, or something else is less important than the essential understanding that specific agents ultimately control and allocate resources ... Someone has to invent rules and apply them: no matter what the macro-structure the socio-political processes of intra- and inter-organizational conflict will prevail.

Many of Pahl's postulates regarding the socio-
spatial structure and Dear and Clark's suggestions
in relation to the relative autonomy of the 'local
state' are supported by an analysis of municipal
housing in Britain.   The allocators of scarce urban
resources exert an important influence upon the
socio-spatial structure and generate differential
housing opportunities for individuals in the sector.
The operation of rules, priorities, and allocations
policies effectively limits or expands opportunities
for prospective and existing tenants and in part
determine 'who gets what type of housing when and
where'.   Such rationing mechanisms are not devised
clandestinely in a covert manner but are the result
of the interaction between a number of interest
groups within the local authority working under the
broad statutory requirements set by central government.   In a decentralised system lower levels of
management interpret centralised policy decisions.
In effect this is the interface between the consumers and the allocators for it is here that the
eligibility of the former is decided.   The models
suggested by Pahl are insufficient to stand alone
but nevertheless cannot be precluded from any
analysis of the socio-spatial structure as they take
us closer to understanding the constraints which
affect residential decision-making.   The next step,
however, is to root the institution within the
society of which it is a component part for although
local authorities undoubtedly have been afforded
interpretative latitude there is what Kirby calls
'the ultimate constraint of the state' (1979, p. 35).
As Kirby notes, while his examples illustrate the
individuality of local authorities, 'ultimately the
buck stops (or more properly starts) here.   Some
local government functions are mandatory and the
Rate Support Grant has a role to play in all considerations.' (1979, p. 35)
    In attempting to explain any allocative pattern
and why this pattern is as it is, rather than something else, it is important that qualifications
such as those noted by Kirby are recognised.
Explanation of a particular allocative structure
and the spatial manifestations of its behaviour
require the researcher to move outside the phenomena
under study in order to understand spatial differen-
tiations in their societal context.   This means
that we must recognise the structural constraints
imposed upon individuals and institutions by a
society.   A differentially rewarded occupational
structure imposes constraints upon certain

individuals (e.g. the homeless,single-parent
families) which minimises their choice of type of
accommodation and its location, forcing them into
the least desirable areas.   The spatial distribution of resources such as housing however can either
'reinforce, reflect or reduce the inequalities
engendered by the differentially rewarded occupational structure' (Pahl, 1977b, p. 50).   It is at
this level that a focus upon the institutional
generation of housing opportunities has much to
offer in preference to earlier attempts to explain
the socio-spatial structure.   More general theoretical interpretations, be they politicist or reductionist, are unsatisfactory, because, as yet, they
have failed to account for local variations in state
activity.

Related Work in Other Social Sciences

The development of institutional approaches in
geography has been linked primarily with developments
in sociology, especially urban sociology.   In
other respects, this development has been largely
independent of work in other social sciences.
Geographers in several fields have become aware of
the importance of institutional action from their
own studies, and their work has been little
influenced by people studying institutions in
political science, sociology and economics.   More
exposure to ideas from other disciplines may help
to enlarge the conceptual and theoretical vocabulary
of institutional study in geography.   Some
approaches to the study of institutions and organisations have been cross-disciplinary, and so the
structure of this section is based on a series of
topics some of which cannot be identified entirely
with one discipline.

Administrative Science

Interest in the function and operation of governmental institutions has resulted in the development of
the sub-discipline of administrative science.   Its
concerns include how administrative systems work,
the relative power of different branches of government, the workings of a particular institution, such
as a government agency or a special commission, the
budgeting process, and the adoption and spread of
new ideas.
     An early study of this type, which has become
very influential, was Selznick's study of the

Tennessee Valley Authority (1949).   Selznick discussed the growth and functioning of the Authority,
focusing on the way in which its leaders interpreted
its mission and on the constraints under which it
operated.   His work was of some originality because
it is based on field study of the organisation and
its actual mode of operation rather than merely on
a study of the documents establishing it and describing its official mission.
       Another important concern of administrative
science has been with policy-making in government
agencies (Sharkansky, 1970).   Linked with this is
the issue of budgeting and organisational power.
Niskanen (1971) proposed the hypothesis that governmental agencies seek to maximise the budgets they
receive, suggesting also that they are in a strong
bargaining position because they control the information about the programmes they operate.   Meier
(1980) has attempted to evaluate the power of government agencies in relation to their resources and
their autonomy.   Recent work in administrative
science can be found in such journals as
Administration and Society, Administrative Science
Quarterly and Public Choice.

Organisation Theory

Another important tradition of institutional study
is that of organisation theory, an important branch
of sociology, especially in the USA, but also a
subject of study in business and management schools.
Work in this tradition can be traced back to Weber
(1947), and more recent exponents include March
(1965; see also March and Simon, 1958) in management,
and Blau (1974) and Etzioni (1961) in sociology.
Although Weber's interest in organisations included
their effect on their environment, most work in this
tradition has been fundamentally concerned with the
factors affecting organisational structure and
social relations within organisations, rather than
with the effects of their structure.   A partial
exception is the concern of Aldrich (1979) and
others with the effects of organisations on their
environments.   More central themes in organisation
theory include the causes and effects of the shape
of the organisational hierarchy, styles of supervision and their impact on employee performance, the
official and unofficial loci of authority, and the
effects of the degree of overmanning or undermanning
in an organisation.

Although the majority of geographers would probably be more interested in the effects of organisations than in the study of organisations as entities, the latter may be of considerable geographical interest. One way in which this may be true is that organisational structure may be spatially expressed; for example, headquarters may be separate from branch plants or oftices, or subdivisions may be located in different places. The relationships between these parts of an organisation may reflect their spatial separation; it may also affect the degree of autonomy the parts have, the nature of their interaction with people and institutions located near them, and the policies they have. In general, the determinants of actions by people in organisations may include factors related to their position within the organisation as well as their official roles. Lack of awareness of intra-organisational factors may prevent a proper understanding of institutional actions, and perhaps a solution to problems arising from these actions.

Developments in Economics

The tradition of institutional economics has long been an important source of dissent to the dominant school of neoclassical economics, with its emphasis on theory derived from assumptions of market clearance and perfect competition. Institutional economists have tended to emphasise other factors, including the role of social factors in influencing the structure of the economy. Thorstein Veblen is often regarded as the founder of institutional economics, and his followers have included John Commons and Clarence Ayres, among others. An influential economist of recent years with links to the institutionalist school is John K. Galbraith (1967), who has stressed that large organisations have a great deal of control over their economic and political environment. The institutionalist school is characterised by its emphasis on the evolutionary character of economic behaviour, and in particular its dependence on the state of technology. In additon, it has stressed the relationship of the economy to social and cultural institutions of the society. Whereas the neoclassical tradition has regarded the organisation of the economy as independent of the political and cultural features of the society in question, institutionalists regard economic factors as closely related to, and dependent on, these extra-economic issues.

The term 'institutional' does not, however, imply a concern with the operation of formally con- stituted institutions in the sense espoused in this book.   Although some institutionalist economists have discussed these issues, they have not high- lighted them to the extent that the term 'institu- tional' might imply.   Much recent work from this school is published in the <u>Journal of Economic Issues</u>, and Gordon (1980) has published a recent textbook.

Another school in economics which has been of importance recently is post-Keynesian economics (Eichner, 1979).   The post-Keynesians were influenced by the work of Keynes and his circle at Cambridge, especially Kalecki, and have developed their ideas further.   It views the rate of invest- ment as the key determinant of both economic growth and income distribution.   In contrast to the neo- classical model, it describes an economic system fundamentally affected by advanced credit and monetary institutions and it recognises the exist- ence of multinational corporations and national trade unions.   Most work in post-Keynesian economics has been macroeconomic, but Eichner (1975) has attempted to reformulate microeconomic theory in the light of the large corporations that are clearly so influential in modern economies.

Much work in the Marxist-influenced political economy school is also highly relevant to a concern with institutions.   In their concern to trace the development of finance capital and its effects on the well-being of the other classes in society, radical political economists have had much of interest to say about the operation of financial institutions, of large corporations, and of govern- ment.   Many of the analyses and the methods used are of relevance and interest to geographers, whether or not they are working within the theoreti- cal perspectives of this group.

Themes in the Study of Institutions

One of the most important respects in which an insti- tutional approach to geography may show promise of adding explanatory power to other types of approach is the attention which may be given to the princ- iples affecting the behaviour of institutions. Institutions are not neutral bodies which impersonally ally fulfil some function in the economy or society;

they may and do develop their own interests and
policies independent of their ostensible purpose.
These interests may have either of two types of
origin: they may stem from the beliefs or interests
of the people manning the institutions, who use the
institution to advance their own ideas, or they may
stem from an institutional ideology, in which the
institution acts so as to advance its own interests.

Beliefs of Institutional Managers

It is not unreasonable to suppose that managers will
act in the ways they think are appropriate or
right, and there are two forms of legitimation for
these views: the cultural or subcultural values to
which they subscribe, independent of their membership of the institution, and the values inculcated
by virtue of membership. In the first category,
the relevant values may arise from the class membership, upbringing and education of the people concerned, and reflect the values of the society from
which they originate. However, the people likely
to become managers are drawn disproportionately
from certain types of background; their selection
and promotion will have been related to this background. One effect will be to make the managers
aware of, and probably sympathetic to, those of
similar backgrounds. Another effect stems from
the education and training the managers have had; a
training in engineering, in economics, or in archi-
tecture may predispose managers to emphasise certain
types of problem, or to look for certain types of
solution.
     The beliefs and policies of managers may also
be influenced by their recent environment, by the
experience of working within an institution and
absorbing its norms and values. It is clear that
organisations have their own styles and ideals, and
that a successful manager will have been strongly
influenced by them. In some cases, promotion may
be dependent on a junior employee being seen to subscribe to the organisation's values. This may
include a willingness to place the organisation's
good above personal considerations, most likely by
being prepared to change jobs and housing at the
organisation's behest.

Institutional Ideology

The ostensible object of an institution will clearly

guide the way it operates, but institutional styles may develop which are largely independent of the mission. Some institutions try to present a favourable image of themselves, both externally and internally. In some cases, this image may be one of caring, in others one of efficiency. Some organisations develop a reputation for a particular style of interaction with their clients, and this style may be absorbed by or inculcated in their employees. Organisations may also be distinguished for their style of internal operations. One style might be a highly competitive one, where junior staff are encouraged to outdo one another in competition for promotion. An alternative style is co-operative, where employees are encouraged to regard the organisation as a mutually supportive and secure environment. Sometimes this organisational style may be the result of the wishes of senior management, but sometimes it may have evolved over a long period without clearly discernible origin.

Of more direct geographical importance may be the degree of local involvement of an institution. Some institutions may be based in one town or city and, as a matter of policy, concentrate their operations in that place, while others may be national or international in orientation. Among those which operate at the national scale, there may be differences between those where local managers and personnel are expected to become identified with the local community and those where local loyalties are discouraged and are expected not to conflict with loyalties to the institution as a whole.

There has been some work in geography on questions such as these, including work on the attitudes and ideologies of institutional managers (Sewell, 1971). A recent example is Knox and Cullen's study (1981) of British planners, which analyses the social background of senior planners and finds that they tend to share distinctive attitudes about their roles, seeing themselves as mediators, as guardians of the environment, and as managers of the urban system.

Institutional Goals

Institutions can be likened to people in that they seem to develop an independent interest in self-perpetuation and growth apart both from the ostensible role they are intended to fulfil and from the interests of the individuals working within them

33

(Pahl, 1979b). This is not to say that institutions do not direct their activities towards their ostensible mission - although the single-mindedness with which they do so may vary - nor that the individual welfare of their employees is disregarded; however, institutional actions may not be reducible to either of these types of motive.

The tendency for institutions to wish to preserve themselves may be explicable in terms of the motives of individuals, who may well wish to preserve their jobs by preserving the institution, and who may recognise their prospects for promotion as better in the organisation they are already in than they would probably be with a new organisation. Institutional growth, similarly, offers opportunities for people currently in the institution to increase their salary, status or power without damaging the interests of internal competitors, a theme made famous by Parkinson (1957). Steinhart's lively account (1972) of American environmental policy-making shows how increased environmental awareness was used by congressional committees and federal agencies to increase their own size and power.

Individual support of institutional growth may stem from other values held by employees, however. As mentioned earlier, training and experience may lead employees to feel attachment to the welfare of an institution for its own sake. Loyalty to the office or to the department may arise naturally in long-serving employees, and may be actively encouraged by senior management. Respect for institutional traditions and belief in the value of institutional styles may lead employees to work for the success of the institution irrespective of the personal advantages it may bring them.

Research in organisation theory has attempted to identify the conditions which lead to these attitudes and the conditions which lead to reluctance of employees to internalise organisational goals. One important strand of this research is concerned with the effects of different styles of leadership on organisational morale. Another is concerned with the effects of environmental conditions, following from the classic Hawthorne experiments, where pleasantness of working environment was found to be closely related to productivity. Thirdly, under-manning theory suggests that organisational loyalty may be greatest when the number of employees is just below the optimum for the task to be done. In any event, attitudes of employees to their job and to their organisation may affect not only their job

effectiveness, but also their attitudes to their
clients and, where relevant, to the service they
are supposed to be performing.

Review of the Chapters

One of the few geographers who has already argued
for the adoption of an 'institutionalist perspective'
is Stephen Gale (Gale and Atkinson, 1979).   In his
chapter, Gale develops some of these arguments,
drawing in part on his experience with the evaluation of the US Community Development Block Grant
scheme.   The chapter is mainly philosophical in
nature, however, relating his interests in rules
and institutions to the nature of social theory and
action and to science as a question-answering
activity.

Commenting on the small amount of methodological work done on the analysis of the effects of
rules and institutions, Gale sees a need to reconceptualise the nature of theory in geography to
allow for the treatment of these effects.   He
criticises work on residential mobility and on
regionalisation for neglecting these issues.
Government programmes and urban planning may have
direct effects on residential mobility, as in the
case of such US schemes as the Experimental Housing
Allowance Programs, but there may be much more
important indirect effects, for example on the
financial characteristics of rehabilitation investment and, generally, major effects on the perception,
judgements and contexts of actors in the housing
market.   In analysing the role of regional units
in policy implementation and evaluation, Gale draws
attention to the institutional determinants of
regional definition; the regional units for which
data are collected and within which programmes are
set up are governed by sets of rules and are institutionally determined.

Gale suggests that institutional activities
can be divided into three types: agenda setting,
procedural resolution (selection of alternatives
which can be chosen) and decision procedures.
These institutional activities can be further
divided into bargaining, management and rationing,
and can also be classified according to their
temporal extent and sequence.   This typology of
economic activity shows that the bargaining activity
making up most of contemporary microeconomics is
only a small part of one category, repeating only

one type of transaction in one specific institutional context.

In closing, Gale claims that the study of institutions has an intermediate position between one-
dimensional models of behaviour, such as the simplified models of microeconomics, and the three-
dimensional world of the humanists. He sees the
study of institutions and rules as two-dimensional,
more heterogeneous than the former and much less
sophisticated than the latter. His prescription
for geography and the social sciences is to develop
this two-dimensional approach, realising and
accepting that it has neither the advantages nor the
drawbacks of the other perspectives.

The remainder of the chapters are intended to
explore the possibilities of an institutional perspective in various subfields of geography. They
contain elements of review material and reports of
original research in different proportions, and
most authors attempt to generalise about the problems
and potentials involved in an approach of this type.
In the remainder of this introduction, the papers
will be previewed briefly with an emphasis on themes
and findings of interest beyond the specific
empirical context of the paper.

Gordon Clark discusses institutions in rural
and agricultural matters, mainly with reference to
Britain. He directs attention to the problems in
identifying the objectives of government institutions. Policies may have several different objectives, perhaps to appeal to different interest groups.
Changes in government may lead to relatively little
change in agricultural policy, there being more
consensus between the parties, at least in Britain
and the USA, than on most other topics. Problems
may arise from the typically substantial time lag
between identification of a problem and implementation of policy to deal with it, perhaps to the
extent that the policy is out of date by the time it
is introduced. Governments seem to prefer financial
policies even when a lack of money is not the
difficulty, perhaps because grants are popular or
because they are easier to administer than non-
financial measures. They may also consider short-
term effects of a policy rather than its long-term
costs and benefits.

Clark classifies and discusses the principal
methods available to governments for implementing
rural policy. The most direct method, involving
land ownership, is expensive and sometimes unpopular.
Less direct methods include subsidies, development

36

control, provision of infrastructure, and taxation.
Another important method involves the creation of
special institutions charged with specified tasks,
with budgets intended to help promote rural development in various ways.

Lack of clarity is a major feature of policy.
Governments simultaneously operate policies with
several objectives and these policies sometimes
conflict with one another. Measures intended to
accomplish one aim may have detrimental effects on
another. Sometimes the problem may result from
lack of co-operation between different sections
within an institution, each section trying to
achieve some goal without regard for its effect on
the goals of other sections. Conditions of administrative fragmentation and departmental rivalry may
increase the difficulty of co-ordination of policy,
even when such co-ordination is explicitly attempted.

In O'Riordan's chapter, the institutional background to policy is further explored in the context
of environmental quality. He echoes Clark's points
about the way in which results emerge from the
workings of many different arms of government, often
without any clearly identifiable decision being
apparent. He quotes the Outer Circle Policy Unit's
statement (1979) that 'policy is secreted in the
interstices of individual decisions, and evolves
over time as a consequence of the decisions that
are taken and carried out.'

O'Riordan discusses several aspects of policy
related to the environment in both Britain and the
USA, and introduces the concept of the policy-
making culture. This culture differs susbtantially
between the two countries. American policy- making
is based on a written constitution with checks and
balances between the various arms of government,
on a mistrust of officialdom, and on the premise
that interest groups will conduct extensive lobbying
to influence policy. In Britain there is no
written constitution, government appears to act more
as a unit, many decisions are left to the discretion
of officials and civil servants, and policy is
formed through consultation between interested
parties in private. In addition, the relationships
between central and local government are very
different, and, in Britain, national policy must
also take into account that of the European
Community. O'Riordan also contrasts the institutions existing for policy review in the two
countries. These institutions include committees
of legislators, public inquiries, public advisory

bodies, and the courts, and the national differences in importance between these institutions, and the effects of these differences, are indicated.

In conclusion, O'Riordan mentions some of the problems encountered by the student of institutional activities, including the necessity of considering political, economic, administrative and legal factors in studying decision-making and policy formation. In matters of public concern, especially in Britain, the student may have to confront an official strategy of containment, designed to keep negotiations and decisions out of the public eye. O'Riordan recommends the use of case studies, preferably from the perspective of an insider, as by far the best way of getting round this problem.

The following chapter, by Chapman, is more of a case study, in which he examines some of the effects of US environmental legislation on industrial location. His focus is on the geographical effects of one type of institutional action, the passage of laws by the US federal government, but other governmental activities, such as the Environmental Protection Agency's promulgation of guidelines, the courts' decisions on the applicability of environmental laws, and state-level legislation and executive action all affect the way in which the laws are applied. Another important theme is the apparent inconsistency of federal government policies, expressed environmental and energy goals being clearly incompatible.

There may be important contrasts between levels of the administrative hierarchy. Chapman suggests that the way in which federal policy is enforced may be influenced by the enthusiasm, or lack of it, with which local agencies carry out their enforcement duties. This may vary greatly between states according to the likely effect of enforcement on the local economy, and according to the general ideology with respect to environmental ethics and to federal regulations.

Another theme developed by Pahl and relevant here is that of bargaining. The eventual water pollution discharge permit is the result of a bargaining process between companies and government agencies, with the larger companies in a better bargaining position than the small ones. The offset policy in air pollution control introduces scope for several interesting manoeuvres, such as the inflation of emissions by operating old equipment so as to preserve scope for subsequent offsetting, and the creation of a 'market in dirty air' as a means

of evading the barrier to economic growth in non-
attainment areas.   Chapman's demonstration of these
and other 'subtle and unexpected influences' of
environmental policy on industry may well have
counterparts in many other arenas of government
action.

In Chapter 6, Thomas Clark undertakes a wider
discussion of the role of the state in regional
development.   Confining his analysis to the
advanced capitalist countries, he reviews the partial
and somewhat simplistic accounts of state activity
offered by existing theories of regional development.
The neoclassical approach sees the role of government only as manager of the market-place; although
some recent authors have studied its effect in
supplying social capital, state policy is regarded
as an exogenous intervention.   Cumulative causation
theory casts the state in a benevolent role, trying
to counteract the inequalities resulting from cumulative causation, but it is not clear why government
should act in this way.   Although Marxian theory
has much to say about the capitalist state,
writers on unequal exchange theory have tended to
ignore this topic, restricting the state's role to
legitimising the existing system of production and
easing crises of accumulation by mobilising reserves
of labour and capital.

Clark then goes on to develop a synthetic model
of regional development combining insights from
each of the three types of theory reviewed.   This
is done first by recognising the relationships that
different parts of a government may have to
regionally or sectorally based interest groupings.
The political economy of regional development is
seen in terms of two feedback loops.   The inner
loop relates state actions to regional economic
capacities and to regional aspirations, which combine to produce sets of socio-spatial outcomes.
These outcomes influence the power relations between
different regionally-located interest groups, and
create the context for new state policies.   The
outer loop links events and power relations to
institutional change and to the structural transformation of the state, which in turn will affect the
regional development process.

Clark's discussion of regional development
raises several points applicable in other areas.
The consideration of the motives for state action
should be essential to all those engaged in policy-
related research, who may have to consider how
government actions may help or hinder different

interest groups in different regions. The conception of the state filling many different roles is
also an important one, especially when these roles
are in conflict - for example, the state as manager
of the market-place may also be a participant in
the market-place, and the state as arbiter may have
to adjudicate as to whether it has acted properly.
The two-way nature of the relationship between government and country should also be stressed.
Government policies are not just exogenous shocks
whose impact can be studied. Geographical reality
can have an effect on those policies, which may
respond to the nature of a perceived problem, and
to the relative strengths of regional and sectoral
interest groups.

In the following chapter, Flowerdew investigates
internal migration from the point of view of institutional activities. Here government plays a major
part, but attention is given to other large organisations, including multi-locational firms, and how
their structure affects migration. He stresses the
numerical importance of migration associated with
institutional influences, including personnel
transfers and moves associated with promotion or
career development in different professions. The
amount of migration within an organisation will
depend on the spatial structure of organisational
operations, and on its personnel policy. Many
multi-locational organisations have a policy of encouraging employee mobility to foster company
loyalty at the expense of local affiliation.

One important problem in applying findings from
one country to another arises from institutional
differences. In the United States, for example,
most professionals are licensed by individual
states; their migration may be greatly restricted
by difficulties in transferring qualifications from
one state to another. In Britain, the banks have
a large number of branches and are active in transferring employees around the country; in the USA,
however, there are legal restrictions on the geographical extent of a bank's operations, and such
transfers cannot take place.

For most households, migration is dependent on
finding a job and accommodation at the destination.
Individual access to housing is institutionally
determined in large measure, with local authority
housing officers and building society managers acting
as gatekeepers. Similarly, company personnel
managers and employment agency officials could be
regarded as gatekeepers to employment opportunities,

with employer demands for previous experience or
paper qualifications acting as criteria for access.
Another important institutional factor related to
migration is the way in which vacancies are
advertised.   People in executive and professional
employment tend to move longer distances on changing
job, partly because their jobs are advertised on a
national or international basis, while unskilled
manual jobs are usually advertised only in a local
newspaper, or perhaps even by word of mouth.

The following chapter brings us to the study of
housing: as mentioned earlier, one of the areas in
which the institutional approach in geography has
been most fully developed.   Perhaps it is not surprising, therefore, that Williams gives particular
attention to theoretical issues in the context of
his discussion of building society operations in
Dudley.   He begins by reviewing the criticisms made
of the early managerialist approach from the standpoint of Marxist political economy.   Agreeing with
the importance of studying the relationship of
scarcity and inequality to the capitalist system, he
nevertheless argues strongly that a focus on
managers and institutions has much validity.   The
processes of capital accumulation and class struggle
are obviously important, but the study of institutions
provides an opportunity to find out when and how
these take material form.

Williams concludes that the lending policies of
building societies will differ according to the
competitive positions of the societies.   They will
be more or less conservative, and will or will not
lend on marginal properties, according to their
local and national positions.   If it is trying to
establish a footing in a new market, for example,
a society may be prepared to take risks unacceptable
to one with a better established position.   There
are differences in lending patterns, therefore,
between building societies; their control by outside
forces is not absolute (and indeed much of this control is exerted by government financial policy, and
can be claimed to be institutional as well as
structural).   Managerial discretion is limited, but
it does exist.

He offers two conclusions applicable for the
whole of geography, not just the study of housing.
One is that more attention must be directed towards
conceptual frameworks and the way they shape our
analyses.   The other is that conventional survey
research may be of limited value in the study of
questions related to ideology and politics, and

interviews and participant observation may be
better suited to uncover the operations of discreet
or sensitive institutions.

In Chapter 9, Johnston writes on the housing
and social composition of American suburbia as it
has been affected by two facets of governmental
organisation, reflecting the Constitutional separation of powers between federal and local government,
and between executive, legislative and judiciary
branches of government. The relatively large
amount of local control is an important factor in
creating and sustaining income and class disparities
between affluent suburbs, which have been able to
zone out low-cost housing and unattractive industrial and commercial development, and an increasingly poor central city, which is left with the land
uses and people excluded by the suburbs, a high need
for social service provision and a low capacity to
pay for it (Cox, 1973).

The main body of Johnston's paper is a discussion of how Supreme Court decisions have or have
not restricted the autonomy of municipalities. He
discusses how these decisions have affected the
access of social and racial groups to suburbia, and
also how these decisions themselves may be related
to the ideologies of the Supreme Court judges.
These judges can be regarded as institutional
managers par excellence. Just as a building
society manager has discretion in how he or she
interprets a given set of institutional policies, so
the justices have discretion in how they interpret
the Constitution. Just as the building society
manager is appointed because he or she is expected
to carry out the policies as the national headquarters would wish, so a Supreme Court justice is
appointed because he or she is expected to interpret
the Constitution as the President would wish.
Their power and influence, of course, are much
greater than those of Pahl's 'urban managers', their
contact with the public is minimal, and it is
difficult to think of them as 'gatekeepers' or
'middle-dogs'. However, they are clearly inter-
preters not makers of policy, and they have discretion within constraints imposed by the Constitution.

A theme in this chapter which may be very
important in many other contexts is the interaction
of institutions, in this case the Supreme Court and
local governments. These two institutions are unequal, however, the Supreme Court (and the other
courts) acting only to uphold or to dismiss
challenges to the actions of local governments.

The Supreme Court can then be regarded as putting
constraints on the autonomy of local government.
In the context of the chapter, these constraints in
turn affect the kinds of constraints local government can put on individual property owners,
developers and citizens. The ability or inability
of local governments to impose certain constraints,
such as zoning ordinances, can be directly related
to the resulting economic, social and political
geography of suburban municipalities and the metropolitan areas in which they are located.
     The focus of Chapter 10 returns to Britain,
where Kirby discusses a number of issues concerned
with education. He begins by reviewing the
evidence for regional and local variations in educational provision and attainment, first at the
national and then at the local scale. He then considers the question of why these variations exist.
One possible explanation lies in the spatial organisation of school systems, and Kirby shows that the
apparently sensible and administratively convenient
principle of the 'neighbourhood school' results in
social polarisation in a city which is socially
segregated. This is one example of many situations
where a policy that can reasonably be claimed to be
'obvious' or administratively convenient may lead to
intensified segregation or inequality.
     He proceeds to examine the professional norms
of local education managers as reflected in recent
publications, an enterprise clearly within the
managerialist tradition. Here he finds a lack of
statements about the purpose of the education
system and 'only a limited interest' in non-
technical issues (like comprehensive education or
core curricula), with concepts like 'efficiency'
and 'planning' given higher priority than the
achievement of social aims.
     Kirby also devotes his attention to linking
work on the administrative control of local events
to the theory of the state in late capitalism.
Like other contributors, he employs the concept of
the local state, stressing three important aspects:
the relations of the local state to the local
economy, to local political demands, and to the
national state. On the first point, Kirby claims
that education is financed in part to fulfil local
needs, with local differences in attainment and in
types of education responsive to the local economy.
He also examines the effect of political struggles
on local educational policy, stressing the greater
degree of success likely for middle-class pressure

Tom Manion and Robin Flowerdew

groups, and the tendency of working-class action
groups to focus on peripheral questions (see the
related discussion on neighbourhood politics by Ley,
1974). On the issue of central-local relations,
Kirby quotes the standard view 'that educational
policy has owed little to central control', but
indicates that the national state is now asserting
greater financial control. Perhaps two points may
be made in this context: the conventional wisdom
about the degree of local autonomy may not accord
with reality; and a central authority may choose not
to exert its control over a local authority,
suggesting an illusory degree of autonomy, until
either local independence of action steps outside
acceptable bounds or a national policy change
requires local authorities to make accordant changes.

References

Aldrich, H.E. (1979) Organizations and Environments,
    Prentice-Hall, Englewood Cliffs, New Jersey
Bassett, K. and Short, J. (1980) Housing and
    Residential Structure: Alternative Approaches,
    Routledge and Kegan Paul, London
Bennett, R.J. (1980) The Geography of Public Finance:
    Welfare under Fiscal Federalism and Local
    Government Finance, Methuen, London
Blau, P.M. (1974) On the Nature of Organizations,
    Wiley, New York
Boaden, N. and Alford, R.R. (1969) 'Sources of
    Diversity in English Local Government
    Decisions', Public Administration, 47, 203-23
Broadbent, T.A. (1977) Planning and Profit in the
    Urban Economy, Methuen, London
Byrne, D.S. and Davis, R.L. (1974) 'The Saga of
    Ropery Banks', paper presented at the
    Newcastle-upon-Tyne conference of the Social
    Geography Study Group, Institute of British
    Geographers
Clark, G.L. (1981) 'Law, the State, and the Spatial
    Integration of the United States', Environment
    and Planning A, 13, 1197-228
Cockburn, C. (1977) The Local State, Pluto Press,
    London
Coppock, J.T. (1974) 'Geography and Public Policy:
    Challenges, Opportunities and Implications',
    Transactions, Institute of British Geographers,
    63, 1-16
Corina, L. (1976) Housing Allocation Policy and its

Effects: A Case Study from Oldham CDP, Papers
In Community Studies No. 7, Department of
Social Administration and Social Work,
University of York

Cox, K.R. (1973) Conflict, Power and Politics in the
City, McGraw-Hill, New York

Dahl, R.A. (1961) Who Governs? Democracy and Power
in an American City, Yale University Press,
New Haven, Connecticut

Dahl, R.A. (1963) Modern Political Analysis,
Prentice-Hall, Englewood Cliffs, New Jersey

Davies, J.G. (1972) The Evangelistic Bureaucrat: A
Study of a Planning Exercise in Newcastle upon
Tyne, Tavistock Press, London

Davies, J.G. and Taylor, J. (1970) 'Race, Community
and No Conflict', New Society, 16, 67-9

Dear, M. (1981) 'The State: A Research Agenda',
Environment and Planning A, 13, 1191-6

Dear, M. and Clark, G.L. (1978) 'The State and
Geographic   Process: A Critical Review',
Environment and Planning A, 10, 173-83

Dear, M. and Clark G.L. (1981) 'Dimensions of Local
State Autonomy', Environment and Planning A,
13, 1277-94

Dennis, N. (1970) People and Planning: The Sociology
of Housing in Sunderland, Faber and Faber,
London

Department of the Environment (1977) Housing Policy:
A Consultative Document, Cmnd. 6851, Her
Majesty's Stationery Office, London

Duncan, J.S. (1980) 'The Superorganic in American
Cultural Geography', Annals of the Association
of American Geographers, 70, 181-98

Eichner, A.S. (1975) The Megacorp and Oligopoly:
Micro Foundations of Macro Dynamics, Cambridge
University Press, Cambridge

Eichner, A.S. (ed.) (1979) A Guide to Post-Keynesian
Economics, M.E. Sharpe, White Plains, New York

English, J. (1976) 'Housing Allocation and a
Deprived Scottish Estate', Urban Studies, 13,
319-23

Etzioni, A. (ed.) (1961) Complex Organizations,
Holt, Rinehart and Winston, New York

Fielding, G.J. (1964) 'The Los Angeles Milkshed: A
Study of the Political Factor in Agriculture',
Geographical Review, 54, 1-12

Fielding, G.J. (1965) 'The Role of Government in
New Zealand Wheat Growing', Annals of the
Association of American Geographers, 55, 87-97

Folke, S. (1972) 'Why a Radical Geography Must be
Marxist', Antipode, 4, 13-18

Tom Manion and Robin Flowerdew

Form, W.H. (1954) 'The Place of Social Structure in the Determination of Land Use', Social Forces, 32, 317-23

Galbraith, J.K. (1967) The New Industrial State, Houghton Mifflin, Boston

Gale, S. and Atkinson, M. (1979) 'Toward an Institutionalist Perspective on Regional Science: An Approach via the Regionalization Question', Papers, Regional Science Association, 43, 59-82

Gerstl, J.E. and Hutton, S.D. (1966) Engineers: The Anatomy of a Profession, Tavistock Press, London

Glyn, A. and Sutcliffe, R.B. (1972) British Capitalism, Workers and the Profits Squeeze, Penguin Books, Harmondsworth, Middlesex

Gordon, W.C. (1980) Institutional Economics: The Changing System, University of Texas Press, Austin, Texas

Gough, I. (1975) 'State Expenditure in Advanced Capitalism', New Left Review, 92, 53-92

Grant, M. (1976) Local Authority Housing: Law, Policy and Practice in Hampshire, Hampshire Legal Action Group, Southampton

Gray, F. (1975) 'Non-explanation in Urban Geography', Area, 7, 228-35

Gray, F. (1976) 'Selection and Allocation in Council Housing', Transactions, Institute of British Geographers, New Series, 1, 34-46

Green, L.P. (1959) Provincial Metropolis: The Future of Local Government in South-east Lancashire, Allen and Unwin, London

Haddon, R.F. (1970) 'A Minority in a Welfare State Society: Location of West Indians in the London Housing Market', New Atlantis, 2, 80-133

Harloe, M. (ed.) (1975) Proceedings of the Conference on Urban Change and Conflict, Conference Paper No. 14, Centre for Environmental Studies, London

Harloe, M., Issacharoff, R. and Minns, R. (1974) The Organization of Housing: Public and Private Enterprise in London, Heinemann, London

Harvey, D. (1976) 'The Marxian Theory of the State', Antipode, 8, 80-98

Hatch, J.C.S. (1973) 'Estate Agents as Urban Gatekeepers', paper presented at the meeting of the Urban Sociology Group, British Sociological Association

Hofstadter, R. (1966) The Paranoid Style in American Politics, Jonathan Cape, London

Holloway, J. and Picciotto, S. (eds.) (1978) State and Capital - A Marxist Debate, Edward Arnold,

London

Jackson, P.W. (1976) Local Government, 3rd edn,
Butterworth, London

Johnston, R.J. (1979) Geography and Geographers:
Anglo-American Human Geography since 1945,
Edward Arnold, London

Kirby, A.M. (1979) 'Towards an Understanding of the
Local State', Geographical Papers No. 70,
Department of Geography, University of Reading

Klein, R., Barnes, J., Buxton, M. and Craven, E.
(1974) Social Policy and Public Expenditure,
1974 Centre for Studies in Social Policy,
London

Knox, P. and Cullen, J. (1981) 'Planners as Urban
Managers: An Exploration of the Attitudes and
Self-Image of Senior British Planners',
Environment and Planning A, 13, 885-98

Lamarche, F. (1976) 'Property Development and the
Economic Foundations of the Urban Question' in
C.G. Pickvance (ed.), Urban Sociology: Critical
Essays, Tavistock Press, London, pp. 85-118

Lambert, J.R. (1970) 'The Management of Minorities',
New Atlantis, 2, 49-79

Lambert, J.R., Blackaby, R. and Paris, C. (1975)
'Neighbourhood Politics and Housing Opportuni-
ties' in M. Harloe (ed.), Proceedings of the
Conference on Urban Change and Conflict,
Conference Paper No. 14, Centre for Environmen-
tal Studies, London, pp. 167-99

Lambert, J.R., Paris, C. and Blackaby, R. (1978)
Housing Policy and the State - Allocation,
Access and Control, Macmillan, London

Large, D.C. (1957) 'Cotton in the San Joaquin
Valley: A Study of Government in Agriculture'
Geographical Review, 47, 365-80

Lasswell, H.D. (1958) Politics: Who Gets What, When,
How, Meridian, Cleveland

Leonard, S. (1979) 'Managerialism, Managers and
Self-management', Area, 11, 87-8

Ley, D. (1974) 'Problems of Co-optation and Idolatry
in the Community Group' in D. Ley (ed.),
Community Participation and the Spatial Order
of the City, British Columbia Geographical
Series No. 19, Tantalus Research, Vancouver,
pp. 75-88

Lipset, S.M. (ed.) (1969) Politics and the Social
Sciences, Oxford University Press, New York

Lloyd, P. and Dicken, P. (1977) Location in Space:
A Theoretical Approach to Economic Geography,
2nd edn, Harper and Row, London

Manners, G. (1962) 'Regional Protection: A Factor

in Economic Geography', Economic Geography,
38, 122-9

March, J.G. (ed.) (1965) Handbook of Organizations,
Rand McNally, Chicago

March, J.G. and Simon, H.A. (1958) Organizations,
Wiley, New York

McCallum, J.D. and Adams, J.G.L. (1980) 'Charging
for Countryside Recreation: A Review with
Implications for Scotland', Transactions,
Institute of British Geographers, New Series,
5, 350-68

McDermott, P.J. and Taylor, M.J. (1976) 'Attitudes,
Images, and Location: The Subjective Context
of Decision Making in New Zealand Manufacturing', Economic Geography, 52, 325-47

McKay, D.H. and Cox, A.W. (1979) The Politics of
Urban Change, Croom Helm, London

Meier, K.J. (1980) 'Measuring Organizational Power:
Resources and Autonomy of Government Agencies',
Administration and Society, 12, 357-75

Miliband, R. (1969) The State in Capitalist Society,
Weidenfeld and Nicolson, London

Mitchell, B. (1971) 'Behavioral Aspects of Water
Management: A Paradigm and a Case Study',
Environment and Behavior, 3, 135-53

Nelson, H.J. (1952) 'The Vernon Area, California -
A Study of the Political Factor in Urban
Geography', Annals of the Association of
American Geographers, 42, 177-91

Niner, P. (1975) Local Authority Housing Policy and
Practice - A Case Study Approach, Occasional
Paper No. 31, Centre for Urban and Regional
Studies, University of Birmingham

Niskanen, W. (1971) Bureaucracy and Representative
Government, Aldine, New York

Norman, P. (1975) 'Managerialism: A Review of Recent
Work' in M. Harloe (ed.), Proceedings of the
Conference on Urban Change and Conflict,
Conference Paper No. 14, Centre for Environmen-
tal Studies, London, pp. 62-86

Outer Circle Policy Unit (1979) The Big Public
Inquiry, OCPU, London

Pahl, R.E. (1970) Whose City?, Longman, London

Pahl, R.E. (1975) '"Urban Managerialism" Reconsidered' in R.E. Pahl, Whose City?, 2nd edn,
Penguin Books, Harmondsworth, Middlesex,
pp. 265-87

Pahl, R.E. (1977a) 'Stratification, the Relation
between States and Urban and Regional Development', International Journal of Urban and
Regional Research, 1, 6-18

Pahl, R.E. (1977b) 'Managers, Technical Experts and
    the State: Forms of Mediation, Manipulation
    and Dominance in Urban and Regional Development'
    in M. Harloe (ed.), Captive Cities: Studies in
    the Political Economy of Cities and Regions,
    Wiley, London, pp. 49-60
Pahl, R.E. (1979a) 'A Comment', Area, 11, 88-90
Pahl, R.E. (1979b) 'Socio-political Factors in
    Resource Allocation' in D.T. Herbert and
    D.M. Smith (eds.), Social Problems and the City:
    Geographical Perspectives, Oxford University
    Press, Oxford, pp. 33-46
Paris, C. (1974) 'Urban Renewal in Birmingham,
    England - An Institutional Approach',
    Antipode, 6, 7-15
Park, R.E. (1952) Human Communities: The Collected
    Papers of R.E. Park, Free Press, Glencoe,
    Illinois
Parkinson, C.N. (1957) Parkinson's Law, or the
    Pursuit of Progress, John Murray, London
Pickvance, C.G. (1978) 'Explaining State Interven-
    tion: Some Theoretical and Empirical Considera-
    tions' in M. Harloe (ed.), Proceedings of the
    2nd Conference on Urban Change and Conflict,
    Conference Paper No. 19, Centre for Environmen-
    tal Studies, London, pp. 24-32
Poulantzas, N. (1969) 'The Problem of the Capitalist
    State', New Left Review, 58, 67-78
Prest, A.R. (1978) Intergovernmental Financial
    Relations in the United Kingdom, Research
    Monograph No. 23, Centre for Research on
    Federal Financial Relations, Australian
    National University, Canberra
Randall, V. (1981) 'Housing Policy-making in London
    Boroughs: The Role of Paid Officers', London
    Journal, 7, 161-76
Redcliffe-Maud, Lord and Wood, B. (1974) English
    Local Government Reformed, Oxford University
    Press, London
Rex, J.A. (1968) 'The Sociology of a Zone in
    Transition' in R.E. Pahl (ed.), Readings in
    Urban Sociology, Pergamon Press, Oxford,
    pp. 211-31
Rex, J.A. and Moore, R.S. (1967) Race, Community and
    Conflict - A Study of Sparkbrook, Oxford
    University Press, London
Robson, B.T. (1975) Urban Social Areas, Oxford
    University Press, London
Robson, W.A. (1966) Local Government in Crisis,
    Allen and Unwin, London
Sauer, C.O. (1918) 'Geography and the Gerrymander',

American Political Science Review, 12, 403-26

Schudson, M.S. (1979) 'On Tourism and Modern Culture', American Journal of Sociology, 84, 1249-58

Selznick, P. (1949) TVA and the Grass Roots: A Study in the Sociology of Formal Organization, University of California Press, Berkeley, California

Sewell, W.R.D. (1971) 'Crisis, Conventional Wisdom, and Commitment: A Study of Perceptions and Attitudes of Engineers and Public Health Officials', Environment and Behavior, 3, 23-59

Sharkansky, I. (1970) Public Administration: Policy-making in Government Agencies, Markham, Chicago

Simon, H.A. (1959) 'Theories of Decision-making in Economics and Behavioral Science', American Economic Review, 49, 253-83

Steinhart, J. (1972) 'The Making of Environmental Policy: The First Two Years' in R.B. Ditton and T.L. Goodale (eds.), Environmental Impact Analysis: Philosophy and Methods, University of Wisconsin Sea Grant Program, Madison, Wisconsin, pp. 5-21

Teitz, M.B. (1978) 'Law as a Variable in Urban and Regional Analysis', Papers, Regional Science Association, 41, 29-41

Townroe, P.M. (1971) Industrial Location Decisions: A Study in Management Behaviour, Occasional Paper No. 15, Centre for Urban and Regional Studies, University of Birmingham

Townroe, P.M. (1972) 'Some Behavioural Considerations in the Industrial Location Decision', Regional Studies, 6, 261-72

Wagner, P.L. (1975) 'The Themes of Cultural Geography Rethought', Yearbook, Association of Pacific Coast Geographers, 37, 7-14

Watts, H.D. (1974) 'Spatial Rationalization in Multi-plant Enterprises', Geoforum, 17, 69-76

Weber, M. (1947) The Theory of Social and Economic Organization, Oxford University Press, London

Whittlesey, D. (1935) 'The Impress of Effective Central Authority upon the Landscape', Annals of the Association of American Geographers, 25, 85-97

Williams, P. (1976) 'The Role of Institutions in the Inner London Housing Market: The Case of Islington', Transactions, Institute of British Geographers, New Series, 1, 72-82

Williams, P. (1978) 'Urban Managerialism: A Concept of Relevance?', Area, 10, 236-40

Chapter 2

NOTES ON AN INSTITUTIONALIST APPROACH TO GEOGRAPHY: TWO-DIMENSIONAL MAN IN A TWO-DIMENSIONAL SOCIETY

Stephen Gale

Introduction

Though the issues are hardly new, my recent work on evaluation of the Community Development Block Grant (CDBG) programme for the US Department of Housing and Urban Development (HUD) has raised a number of questions about the utility of the types of theories and methods which we use in geography. I think that it is safe to say that, while we have methods for explaining some classes of behaviour, we as yet have little to help us in the analysis of the effects of rules and institutions. There are, of course, ways of analysing some types of rules in purely formal systems (e.g. Hotelling's linear markets), but, in the main, rules and institutions are simply treated as exogenous contextual variables.

That an institutionally neutral approach makes little sense has been obvious for a long time. At least since Plato, we have been aware of the differential effects of such man-made devices as laws, jurisdictions, and cultural mores. There is now a need to reconceptualise the nature of theory in geography to allow for an explicit and opera-tional treatment of the ways in which rules and institutions influence the design, operation, and impacts of individual and social behaviour.

To a great extent, this essay is derived from several papers I have written on the foundations of the regionalisation problem and the linguistic basis for inference in the social sciences (e.g. Gale, 1977b; Gale and Atkinson, 1979). In this paper, I will repeat the key points of the arguments in the appropriate places, and provide a rather broader account of the changes which I think are needed in the development of theory in geography.

Stephen Gale

The change in perspective that I will propose
is a response to the need to account for the role
of rules, rule settings, and rule-conforming
behaviour.   My motivation is both theoretical and
pragmatic: theoretical in that there is as yet
little in geographic theory which guides the selection and interpretation of alternative sets of
rules; and pragmatic in that the basis for individual and social decisions depends on the ways in
which particular sets and configurations of rules
are specified and implemented.
     Though any one of a number of examples could
be used, I will use the analysis of residential
mobility in relation to the CDBG programme to illustrate some of my remarks.   (As with questions of
industrial and public facility location, transportation network design, and other related issues, this
problem not only is essentially spatial in character
but also has a strongly institutionalised character
in its influences and outcomes.)   I will also draw
on some ideas concerning the regionalisation problem
to provide further motivation.   Following these
comments, I will turn to a brief explication of a
proposal for an institutionalist perspective on
social science and social practice.   Though limited
in scope, the argument will attempt to demonstrate
that some form of institutionalism can be shown to
be part of the tacit structure of our current
theories and data as well as the ethical grounds of
beliefs and actions.
     Even though the purpose of the argument here
may appear to be abstruse, it is actually quite
hard-headed.   If social understanding and praxis
are to be anything more than ambiguously related
forms of intellectual sculpture and criticism, if
there is to be an integrated approach to the understanding and design of programmes for social change
which can offer insight into planning and practice,
inquiry in geography must offer more than descriptions and explanations of patterns of past behaviour
or simplistic normative programmes based on
idealised optimisation or ethical selection rules.
Rather, it must adopt a heterodox perspective which
integrates: (1) descriptions and explanations; (2)
projections of potential consequences; (3) prescriptive and normative judgements; (4) operational
procedures which translate prescriptive analyses
into practice; and (5) experimental modes of valida-
tion and evaluation.   My claim is that this
perspective must be based, at least in part, on a
general understanding of the role of institutions

and institutional rules in different situations and the development of a methodology which allows for explicit consideration of the impacts of alternative rules.

Indeed, the general claim is very simple: institutional rules are two-dimensional constructs which are specifically designed to treat social interaction and behaviour as if individuals and groups were also two-dimensional.

Motivations

Two types of motivation, one epistemological and one substantive, will be used to provide an intellectual setting for the institutionalist argument. Neither motivation should be construed as part of a knock-down proof that institutionalism plays a central role in geographic thought. Rather, they are the issues which influenced me; any number of others might, of course, have done the job just as easily.

Though current social science lore has emphasised an epistemology based on the development and testing of analytic models for explanation and prediction (in an effort to discern social laws), there have recently been strong counter-arguments to the effect that these goals are, if not misplaced, certainly limited in scope (a few examples, from several different fields, include Churchman, 1971; Gale, 1972; Georgescu-Roegen, 1971; Habermas, 1972; and Olsson, 1975). To be sure, some of these claims have been ideologically based, but they have at least provided a methodological conscience to what often are regarded as philosophical truisms. Basically, however, a gap exists between two pers-pectives: on one side there is a tradition of scientific orthodoxy based on analytically motivated forms of model-building using mathematical criteria of description and inference; and, on the other side, there is a heterodox dialectical tradition (often) combining historical, philosophical and critical methods. The first is based largely on extensions of physical science paradigms while the second appears to be grounded in predominantly humanistic perspectives and ideals.

Contemporary social research programmes appear to be clustered on either side of this conceptual gap. The analytic, models-oriented, theory-testing crowd is on one side proclaiming that the true path has been found; the humanists, on the other side, have sought mainly to preserve the integrity of less

Stephen Gale

formalised reasoning methods, to remind us that
even truth is relative, and that there are important
classes of questions that have been ignored by their
opposite numbers. In a sense, it is a kind of
dialectic in which there are neither commonly recognised grounds of dispute (i.e. in terms of the
delineation and character of the conceptual gap),
nor even a reasonably good idea of what kinds of
arguments would resolve them.

What appears to have been ignored in this
intellectual face-off is that there are more pragmatic conceptions of inquiry which can (potentially)
give some structure to the ground between these
seemingly disparate perspectives - at least insofar
as they can delineate approaches which are appropriate for specific classes of questions. For
example, I have recently written a paper on the
logic of questions and answers and its relation to
inquiry in science (Gale, 1977b; this paper refers
to other work on the topic). The argument is
quite simple and is one of the bases for my proposed
institutionalist perspective. If we view science
as a general question-answering process then, by
virtue of the classes of questions which are asked
(e.g. about formation of classes of entities,
definitions and measurements, explanations, value-
based problems, rule conformity), different methods
of analysis and argument can be delineated for each
such class. In this view, the gap between the
positions described above arises not so much from
'in principle' differences, as from heterogeneity of
methodology and the need for an understanding of the
appropriateness of specific strategies of inquiry in
specific circumstances. In other words, these
seemingly different approaches can be considered in
terms of a richer conception of a question-answering
process which does not regard truth and analytic
representation as the only outcomes of inquiry.

Two examples will be used to illustrate the
substantive side of the argument: the analysis of
the effects of government programmes on residential
mobility and the regionalisation question. Each is
intended to provide an indication of the reasons why
a new point of view would be of advantage. Indeed,
the argument is simply an analogy to Nelson's (1981)
untangling of the concepts and principles underlying
the ethos of private enterprise: in this case,
however, the untangling reveals another class of
influences rather than simple confusion.

Residential Mobility

Residential mobility is a conceptually complex but
omnipresent phenomenon.   Though the intellectual
lore of research in the area over the past half
century or so has emphasised the singular efficacy
of explanatory questions such as 'Why do families
move?', it is now becoming increasingly clear that
such enquiries are, if not misplaced, certainly of
limited value.   Of course it is true that the
explanation of mobility is complicated, but so is
the explanation of just about any other social
phenomenon which has economic, demographic, psycho-
logical, geographic and other influences.   And of
course it is true that a general explanation of the
mobility process could help to set the stage for
effective public and private actions.   But, at
least from a pragmatic perspective, what is
interesting and important about the study of mobility
is that it is almost sure to be influenced by the
rules and regulations of social and governmental
relations.   It is at once a cause, an effect, and
an intermediary; localised and spatially extended;
a consequence of public and private policies and a
policy instrument; endogenous and exogenous; and so
on.   Indeed, the focus and contexts of residential
mobility are clearly heterogeneous and as much
dependent on exogenous social structure as on
individual behaviour and choice.

The two decades immediately after World War II
were a time of visionary social science: of a vision
of complete, consistent social theories which would
embody both positive and normative perspectives.
Social psychology, in particular, was pictured as
playing a central role in the development of such
theories.   Questions such as 'Why do families move?'
were effectively transformed into requests for
information and models of individual and group
behaviour in complex settings.   Families, as groups,
were classified by a variety of economic, social,
demographic, and other characteristics; within
particular physical and institutional contexts,
families were then viewed as assessing their place
utility in light of jobs, family needs, housing,
and so on.   The anticipated theory was to have had
two parts: one which assessed how various factors
influenced the choice of place, and a second which
prescribed locational changes in light of comparisons with some normative base.   At the least, the
vision held out an expectation of knowledge; in

principle this knowledge was to be translated into the kinds of prescriptions social scientists like to call 'policies'.

It takes little post hoc vision to see that many of the hopes of these early years were rather ill-founded. The vision of a 'grand theory' was clearly premature; and, perhaps most importantly, the expected translation of the positive into the normative seemed beyond even conceptual reach. Understanding and the grounds for understanding is one thing; but the grounds for translating under-standing into prescription and action was obviously qualitatively different.

Since the planning and policy process involves several different types of questions (i.e. requests for information and decisions), the easiest way to conceive of the heterogeneity of mobility research is by indicating the multiple types of information requirements for answering these questions (Gale, 1976). Types of questions can, in effect, be matched with particular types of information and, barring the development of a comprehensive theory and measurement system, such a typology implies a considerable heterogeneity in the study of residen-tial mobility.

It has perhaps not been apparent from my earlier comments how institutional rules affect the outcomes and impacts of residential mobility. Although some relocations are forced on the household by exogenous conditions (e.g. where displacement is a consequence of redevelopment), most reflect choices among a limited number of available alternatives. The alternatives considered usually correspond to a variety of types of housing. The choice, however, reflects a complex set of conditions acting on housing quality and availability as well as the social and economic situation of the household, and it should be viewed in a more general context. What, for example, are the impacts of specific community development programmes on the housing decisions of individual households? Or, if community development programmes have induced housing adjustments, what subsequent shifts have there been in the housing stock and occupant households?

As it was described earlier, the traditional approach to the analysis of mobility treats the housing adjustment decision as reflective of the family's dissatisfaction with existing consumption patterns - where the 'dissatisfaction' arises from changes in jobs, family conditions, rents, the implementation of specific governmental programmes,

and so on.   Rules, in the form of general regulations or the community development strategies
implemented by individual cities, enter into the
argument in a number of ways: outcomes of decisions
affected by a programme such as CDBG, for example,
form part of the context in which new decisions are
made in subsequent years; in some cases, the actions
are directly controlled by the city's overall
community development programme (such as redevelopment, investment, or code enforcement) or by its
direct effects on the quantity and quality of the
available housing stock; in other instances,
community development programmes indirectly change
the relative attractiveness of housing alternatives
by changing the financial characteristics of rehabilitation investments.   Thus, while community
development activities may not specifically engender
residential mobility as a direct outcome, the rules
of the programmes do materially affect perceptions,
judgements, and contexts for many households, and
may lead indirectly to mobility.

Regionalisation

Another motivation for this institutionalist orientation is the regionalisation problem.   Here
regions are taken as 'units of data collection and
analysis' as well as 'mechanisms for monitoring and
controlling social affairs' (Gale, 1975; 1976; Gale
and Atkinson, 1977; 1979).   At least part of the
core of geography is dependent on the concept of a
region, and almost all theories of spatial behaviour
and organisation use the idea of a region in some
quite fundamental ways.   A quick review of the
literature shows that it is difficult to find a
consistent and well articulated set of discussions
on the idea of a region. (Isard et al., for example,
devote only one relatively long footnote to a discussion of their concept of a region (1969, p.602).
Despite exceptions such as Grigg, 1967, and
Prescott, 1965, however, the general discussion has
been little more than idiosyncratic.)   Equally
significantly, it is difficult to find systematic
series of data which are tabulated for comparable
regions and subregions.   The analysis of urban
areas in the United States, for example, consistently
refers to such institutionally defined units as
Standard Metropolitan Statistical Areas (SMSAs) and
census tracts, but rarely are there similar sets of
data relating to other spatial units.   In regional

planning, on the other hand, we often find that simplified data sets are employed which refer to selected economic characteristics of specific administrative regions, but rarely are there comparable data on the social, political, demographic and environmental characteristics of these same areas (see Moore, 1979, for a more detailed discussion of this issue).

Though it may seem like little more than a truism, it is important to realise that the term 'region' has a variety of meanings, meanings which are associated as much with various cultural, legal, and political traditions as with the nature of the formal properties of spatial partitions. Side-stepping some of the historical considerations of natural, functional, and perceptual regions, it is useful to focus on two conceptions of the term: arithmomorphic regions which are defined by 'crisp' partitions and non-arithmomorphic regions defined by 'fuzzy' partitions. The substance of the regionalisation problem can then be viewed in terms of four related points: (1) that arithmomorphic and non-arithmomorphic views (respectively) give rise to essentially different meanings of 'region', 'boundary', and 'location'; (2) that classical set theory and Zadeh's theory of fuzzy sets can be (respectively) used to represent the concepts of arithmomorphic and non-arithmomorphic regions; (3) that the contrast between the classical and fuzzy set approaches helps to pinpoint the origin of some specific functional problems in spatial systems; and (4) that the possibility of such alternative conceptualisations gives rise to the need for some 'meta-rule' for selecting among the different approaches to regionalisation.

The last point is of critical importance here since it is in such a meta-rule that the roles of institutions and rules take on pragmatic force. At least in a simplified form, there appear to be three questions related to the formation of any meta-rule associated with spatial partitioning: (1) What sets of rules govern the criteria for specifying the types of regional partitions to be considered? (2) What sets of rules govern the choice of the specific type of region to be employed in any particular analysis or decision? and (3) What sets of rules govern the particular configuration of regions to be used? Even though we may only be able to give a rough characterisation of the grounds for answering such questions, it is apparent that these grounds are at least partly institutional rather than purely analytic.

The preceding types of motivation for the development of an institutionalist perspective in geography are two among many. Little has been said, for example, about the use of related constructs in neoclassical economics or the sometimes profoundly suggestive methods for treating price equivalents of housing services (which have only recently been developed for large-scale social experiments such as HUD's Experimental Housing Allowance Program). Indeed, insofar as the articulation of theories about the roles of rules and institutions are concerned, each of these approaches (and many similar strategies) has led to largely ad hoc positions: institutions may provide fixed constraints, prescribe decision rules, and so on, but there is no way of understanding - much less deciding - which of several types of rules or institutions should be employed in any particular case. And even in those rare cases where the only important considerations are those of economic efficiency (and not the overall influence of political, legal, and administrative rules), the problem simply reduces to another form of the meta-rule: that is 'How do we know that the only important considerations are those of economic efficiency?'

The short of it is that there now appear to be important intellectual reasons to begin to develop an institutional perspective on theory in geography - and to invest it with the pragmatic force necessary for the translation of theory into action.

Institutionalism

At the outset, it has to be made clear that the institutionalist perspective which I will discuss here is only one way of approaching the general concern with rules. Marxist analysts, for example, have also urged that social science research programmes be reconceptualised, but in terms which focus on the long-run effects of (tacit) ideological rules on the structure and distribution of resources (e.g. Harvey, 1973; Holland, 1976; and Roweis and Scott, 1978). Others have urged similar reconstructions of theories of decision-making so as to account for a range of specific institutional influences on the behaviour of individuals and groups (e.g. Gale and Moore, 1975). Though often cloaked in the garb of 'anti-neoclassicism', the emphasis of these proposals is far more general:

the issue is whether particular classes of questions, such as those concerning longer-run, aggregate conditions and processes, and the impacts of socio-economic and political institutions, can regain their place in the social sciences in general, and in geography in particular.

Clearly, this is not the place to review and reassess the history and development of social science thought. At the heart of much of the recent work on the methodology of social research, however, are concerns about the relative value of various types and lines of questions. The strong Anglo-American logical positivist tradition based on reductionism and methodological individualism is now being seriously debated (Lear, 1978) and, to my mind, we in geography must begin with equal seriousness to review the motivation and substance of our analyses and their presuppositions.

Institutionalism is little more than a symbol of the roles which social, economic, and cultural rule systems and conventions play. What is of more general importance, however, is the problem of characterising the manner and form in which institutional considerations can be accounted for in geography. When it comes to understanding their broad and diverse roles and impacts, how can we treat the effects of seemingly non-systematic rules in a systematic manner?

Let us consider the problem of characterising the process of planning as a series of institutionally based operations. It is useful to follow Georgescu-Roegen's (1971) lead and use time, in an abstract sense, as the principal conceptual force in the argument. Time imposes limitations of resource availability; it changes a deterministic, stationary world into a stochastic, non-stationary one; it forces us to select our planning horizons and, therefore, the ways in which the future influences the present; and so on. But how do we approach any particular planning or decision-making task? How do we select amongst alternative planning periods and decision-making rules? How do we treat the problem of selecting a means for describing future events? And what are the relationships among description, prescription, and evaluation?

Initially, we must be clear about two points: first, that empirical models and theories are, in some sense, always models of data (Suppes, 1962; Moore and Gale, 1973); and second, that a theory is, most often, an answer to a (very specific and

probably quite limited) question (Toulmin, 1977).
Indeed, social theory and modelling are much broader
and complicated than analogies to explanation-
oriented physical science paradigms would have us
believe (cf. Bromberger, 1966; Korner, 1966, 1976;
Toulmin, 1969; 1977; Gale, 1977a).   Management
strategies and jurisprudence are, for example,
integral to concepts of planning and decision-
making; and, just as institutions influence the
structure of models of data (say, through collection
procedures and the organisation of modes of analytic
representation), so too do institutions affect the
nature of social laws, organisational behaviour, and
individual decisions (Dunn, 1974).

Paraphrasing the Sapir-Whorf thesis, we can say
that 'We see what institutions allow us to see, we
know what institutions allow us to know, and we act
as institutions allow us to act.'   This is, of
course, a very strong claim - which has more than
its share of critics even when restricted to its
normal linguistic context.   At the same time, its
merits are often just as inescapable.   To 'see'
the world of planning problems and decisions, for
example, we can say that we use the 'filters'
shaped by those existing social institutions which
are responsible for the analytical frameworks such
as those connected with information collection and
organisation.   And to use processes of decision-
making, we employ institutional representations of
individual choice and modes of social organisation.
Far from being watered-down Marxism, however, each
of these claims speaks to a pragmatic thesis: that
both the structure of information and social and
individual decisions are simply functions of rules
and institutions.

Ideology notwithstanding, then, the perspective
I am advancing here presupposes a pragmatic approach
to understanding such processes as spatial planning,
an understanding which is based on an explicit
recognition of the influence of the roles of rules
which organise social information and institutions.

Figure 2.1 outlines one way of conceptualising
the sequence of relationships that take place in a
general planning, policy development, and evaluation
process.   Suppose, for example, that as a result of
the implementation of the CDBG programme (say, the
rehabilitation of single and multi-family housing
units) there is a perceived need to develop and/or
reorganise the related social services (e.g.
schools, libraries, bus routes and schedules).
The proposed change in the availability of specific

Figure 2.1: Planning and Decision-making

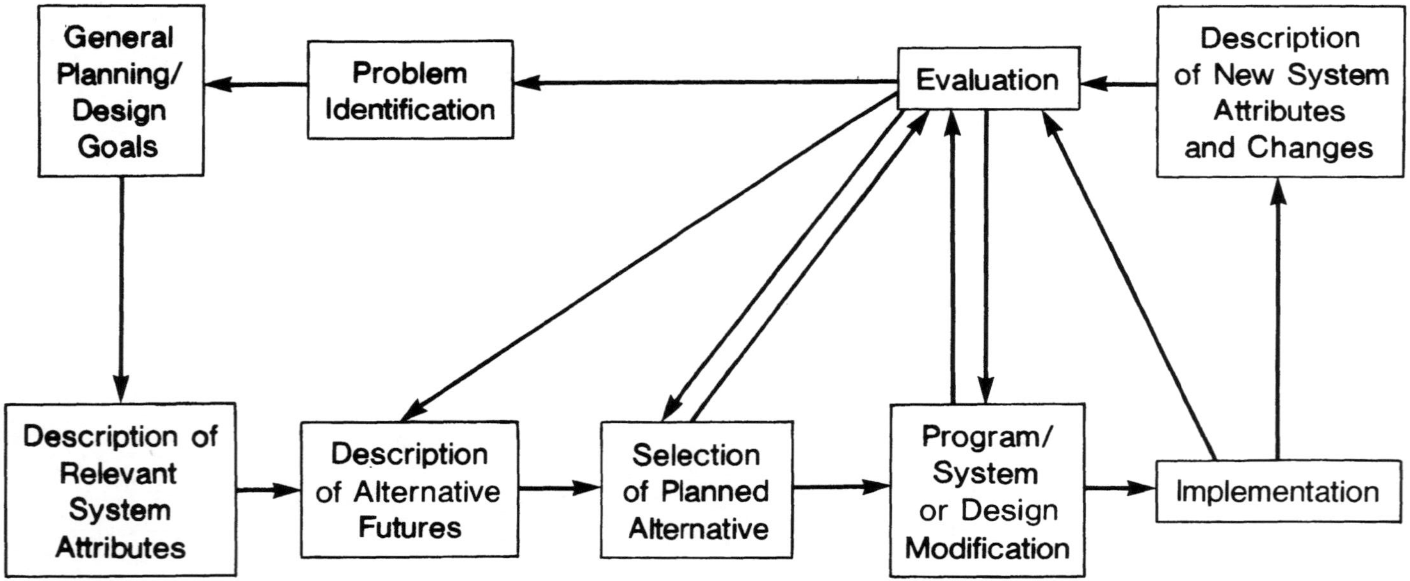

types of dwelling units will obviously affect
specific segments of the population in different
ways and, in turn, will have differential effects
on the composition of the population in the various
sub-areas of the locale. A reasonable (though
complex) question, then, is 'What effects will the
proposed programme have on the population composition and distribution and how will they affect the
distribution, organisation, and provision of
services?'

Traditional approaches to this type of problem
have, for the most part, attempted to separate the
question into a number of parts - usually organised
according to disciplinary interests: one part
addressing, say, the politics of housing provision
and others addressing the relations among population
composition and level, kind of service provision,
and the dynamics of residential mobility. Side-
stepping the knotty issue of problem fragmentation,
it is useful to consider some of the ways in which
the investigation of this type of question might be
treated in a typical programme of analysis.

1. Planning and design goals are treated as
   either: (a) optimisation schemes based on
   analytical representations of preferences
   for housing (with prespecified budget constraints) and a prescribed social welfare
   allocation scheme; or (b) extrapolations of
   assignment and selection procedures which
   are identified through comparisons with
   prior instances of analogous programmes.
2. The identification of the relevant system
   attributes is normally determined solely
   by prescribed analytic models for: (a)
   information organisation; (b) the use of
   surveys to elicit more detailed data on
   residential preferences, family structure,
   income, and so on; and (c) the construction
   of theoretical indicators which conform to
   specific, idealised decision procedures
   (e.g. utility for accessibility).
3. The alternative futures are those distributions of population which are derived
   from either: (a) analytically based
   economic models of investment in housing
   services; or (b) extrapolations of existing
   (or analogous) patterns of selection.

In general, little explicit prior attention is given
to how 'planning alternatives' are enumerated or to

questions of programme design, implementation, and evaluation. An explicit separability is also presumed which relegates planning problems either to a programme which synthesises existing data within the framework of specific types of analytic models or involves the application of specific programme rules. Analytic methods and theories are, in effect, taken as primary and the results of enquiry follow directly from their presuppositions.

A related and perhaps more fundamental problem concerning the role of rule systems also arises in connection with the general orientation of schematic representations such as that illustrated by Figure 2.1: that is, by virtue of their design, 'flow chart' diagrams tend to focus on the boxes rather than the arrows (i.e. the processes by which the observations and activities are carried out).

Consider, for example, the arrow which connects the boxes labelled 'description of alternative futures' and 'selection of planned alternative'. The intuition is that once we are given some kind of list of possible futures (as states and/or as actions), some procedure is used for determining which item (or items) on the list will be treated as the selected alternative (i.e. for subsequent examination or implementation). If, for instance, the procedure is treated as a purely technical issue (say, as part of a market-based exchange paradigm) and subject to a particular type of mathematical characterisation, we would expect that the argument would be based on the use of some form of optimisation strategy. Or, if the procedure is regarded in essentially political terms, some institutionalised form of voting may be thought of as the decision procedure. Underlying any such procedure, however, is an explicit recognition that the 'arrows' - the rule-based relationships - must, themselves, constitute part of the design and selection of the overall strategy and that the analytical structure of the information base must be, in turn, designed to facilitate the use of alternative procedural criteria.

The point here thus reinforces and extends my earlier comments: not only has contemporary social science tended to treat the analytic aspects of methodology as the principal means for identifying and approaching problems, but the selection of a methodology has itself been isolated from considerations of the nature of the type of argument. Broad, complicated issues such as planning thus become translated into problems of analytical

representation and the manipulation of data - both
of which are several steps removed from such
important considerations as data classification and
organisation, and the institutional procedures
which employ the results as part of the larger
planning and policy-making process.   The microcosm
of analytically based explanatory theories thus
becomes the macrocosm of social practice: theory and
practice are linked, almost by fiat, with little
regard to the substance of the links or the type of
chain which has been forged.

Extending this example a bit further, consider
the central role which the box labelled 'evaluation'
plays in Figure 2.1.   Evaluation, in its broadest
sense, is a generic term for a monitoring and comparison procedure within a general learning system
(e.g. a system concerned with assessing impacts and
guiding changes).   Where the veracity of a causal
model is under scrutiny, for example, the evaluation
procedure may be something on the order of a
Popperian falsificationist paradigm or a Carnapian
inductivist principle.   Or where the rules for
selection of a particular plan are at issue, the
evaluation scheme might consist of a system for comparing alternative (numerical and ethical) choice
procedures.   The point here, however, is not that
there is (or can be) a unified theory and methodology
of evaluation, but that the nature of data, models,
and institutionalised decision procedures must
explicitly address some relevant monitoring and
learning scheme.

Unfortunately, there are pragmatic tensions
involved in describing the planning process in
integrated terms.   Social institutions, for example,
often have vested interests in using certain forms
of analytical representations and programme types.
It is inevitable, then,that if we attempt to
develop analytical representations which are
sufficiently flexible to be used as the basis for
evaluation, we will be led to reformulate these
same problems and, therefore, to modify the institutions which we employ to deal with them.   Such
changes are likely, in turn, to generate a shift in
the distribution of economic and political power as
well.

If, however, this type of problem is characteristic of an important class of questions, then the
consequences of failing to recognise the need to
shift the emphases of our theories may be even more
disturbing.   Programmes designed with little regard
to the institutional diversity of the processes of

Stephen Gale

decision-making are likely to be inflexible and generally of little value. (Often, for example, they generate unexpected - or even expected - wind-fall profits for certain groups at the expense of others who fall progressively further behind in their ability to obtain adequate standards of educa-tion, housing, health or other social goods and amenities.) Of course we are always faced with problems of allocating limited resources within a particular value structure and political framework. It should be clear, however, that from the point of view of the means of enquiry there is really no justification for basing decisions on a monotheistic view of analytical relationships and transactions.

Although it has been claimed that rules and institutions play a critical role in spatial planning and related activities, it is still unclear how they can be systematically treated. This is no small order, particularly since there is little systematic information about complex institutional forms and their effects to parallel the ideologi-cally based perspectives represented, say, by the mechanistic picture of self-interested economic entities employed by neoclassical economics or the holistic picture of material/historical change offered by Marxists. In fact, what little knowledge we have about the nature and effects of the structure and organisation of social institu-tions comes either from the relatively idiosyncratic pictures of the behaviour of economic institutions presented by such writers as Commons (1924; 1950) or from general analyses of the behaviour of specific organisations (e.g. March and Simon, 1958; Gale and Moore, 1975; Allison, 1969). These treatments are, however, insufficient to provide a general charac-terisation of the effects of institutions such that a clearly articulated perspective could be advanced. Institutions are treated as isolated, exogenous constraints with little explicit bearing on either the nature of the decision-making or the specifica-tion of the factors in any such decisions. In effect, the existing approaches do not appear to do real justice to the concept of an 'institution' and the ways in which institutions interact and affect broad classes of social decisions.

What seems to be required are theories which allow us to understand and draw inferences about the role of alternative institutions and institu-tional mixes - such as market systems, political systems, administrative systems, community and cultural systems - and the ways in which these

66

institutions organise the activities and agenda which set decision priorities. Markets and the specific rules of market transactions, for example, affect the relative desirability of land in urban areas; community rules and the rules of local social behaviour affect the territorial cohesion of specific groups; administrative procedures set the rules for land use, the priorities for changes in classification, and the structure of the information systems on which decisions are predicated; and legal institutions set the rules governing the use of other institutional rule systems and the rules by which they are enforced and changed. The rules are complicated, the rules about rules are more complicated, and the interactions among rule systems even more so.

On a very general level, it is possible to speak of the roles of institutions and rules in terms of three types of activities:

1. Agenda setting: What rules determine the issues to be resolved?
2. Procedural resolution: What rules set the alternatives which can be chosen?
3. Decision procedures: What rules make decisions concrete and final?

Indeed, the rules governing the process of answering these questions are themselves explicitly dependent on institutional structures.

But speaking of rules is still removed from the more fundamental problem of providing a typology of rule-oriented behaviour and connecting it to specific classes of decisions. Commons (1950) offers such a typology which, although still quite rudimentary, goes some way toward clarifying the linkages between methodological issues and substantive questions concerning the nature and impacts of institutions. The argument consists of two parts:

1. that exchange and allocation decisions are based on three simple types of transaction: bargaining, management, and rationing; and
2. that the temporal extent and sequence of transactions specifies the type of institutional form.

In principle, quite simple: an interaction between the timing of a decision and the kinds of transactions is resolved as a typology of kinds of behaviour. The implications are far-reaching,

Table 2.1: Scope of Transactions

Kinds of Temporal Interaction	Kinds of Transaction			Kinds of Analysis
	Bargaining	Managerial	Rationing	
Negotiation	persuasion, coercion	command	command	Qualitative
Commitments for future action	performance and payment	production of wealth	distribution of wealth	Qualitative
Execution of the commitments	prices and quantities	input and output	budgeting, taxation, wage/price fixing	Quantitative

Source: adapted from Commons (1950), p. 57

however, as examination of Table 2.1 will show.
Notice, for example, that virtually the whole of
contemporary microeconomics is part of only one cell
of the typology - the lower left - and that the
entire picture of transactions depends strongly on
agenda which are set in ways exogenous to the
selection of any particular approach.   The process
of the transaction, moreover, is entirely quantitative as is the process of deciding on the timing to
be considered.   In effect, as Commons would have
it, the entire non-institutional perspective
currently in vogue in much of the analytically
oriented social sciences is only a small part of
what is required to understand and influence
decisions - and, perhaps more importantly, the
methods appropriate to the analysis of economic
behaviour are not generally applicable to the
analysis of other forms of behaviour in different
institutional situations.
     Finally, it can be seen that Commons's form of
institutionalism also provides a constructive background for the rule-oriented approach.   Not only
does it give an explicit typology which accounts for
the roles of rules in alternative settings, but in
addition it sets up a framework for the specifica-
tion of the associated institutional forms that can
be used to formulate and enforce rules: the legal
system (through contractual law) for issues relating
to the distribution of wealth; administrative
committees for issues relating to orders; and so on.

Conclusions

Commons's picture of ideological and methodological
heterogeneity is one part of the argument which
motivated this essay: where a variety of institutions and rules are part of a system of decisions,
then they must be accounted for as such.   The
implications of the argument are thus quite straightforward:

     1. there is a de-emphasis on a unified neo--
         classical economics approach to decision-
         making;
     2. there is an emphasis on the monitoring of
         specific transactions in specific contexts;
     3. there is an emphasis on accounting
         explicitly for the operation of rules in
         these various contexts; and
     4. there is an emphasis on understanding the

Stephen Gale

> ways in which social institutions use
> systems of rules to effect particular
> classes of transactions.

In a sense, it is as if to say that single-
dimensional models of behaviour - and their emphasis
on understanding individual behaviour - are just not
all that critical in a world of rules and institutions.

This brings me to the subtitle of the essay.
In a report of conversation between a reporter and
Frederick Forsyth (the author of espionage stories)
in World Press Review (1980) the following inter-
change is noted:

> Reporter   How do you see yourself in relation
> to bestselling authors such as Graham Greene or
> John le Carré?
> Forsyth   Don't compare me with Graham Greene.
> He creates characters. My books are reporting on
> a grand scale, without style. I write in chronolo-
> gical order without flashbacks. All my characters
> are two-dimensional, like cartoon characters. For
> me, characters and dialogue are there only to tell a
> story.

And so too, in a sense, are institutions and rules.
They set broad plot lines, shape the dialogue, and
set the agenda. The institutions which regulate
social systems are, by design, more heterogeneous
than single-dimensional models; at the same time,
they are far too blunt to take the so-called multi-
dimensional world of humanism seriously. Institutions and rules are, in short, designed to treat
cartoon-like images of the world. The primary
problem for the social sciences is to learn how to
use them more effectively, not to rail against their
avowed heterogeneity or lack of sophistication.

Acknowledgement

An earlier draft of this paper was presented at the
First World Regional Science Congress, Cambridge,
Massachusetts, June 17, 1980. The work that provided the basis for this publication was supported,
in part, by funding under a Co-operative Agreement
with the US Department of Housing and Urban Development (Grant No. HUD-H-2887-CA). The substance and
findings of that work are dedicated to the public.
The author and publisher are solely responsible for

the accuracy of the statements and interpretations
contained in this publication. Such interpretations do not necessarily reflect the views of the
Government.

References

Allison, G. (1969) 'Conceptual Models and the Cuban
Missile Crisis', American Political Science
Review, 63, 689-718

Bromberger, S. (1966) 'Why-Questions', in
R.G. Colodny (ed.), Mind and Cosmos,
University of Pittsburgh Press, Pittsburgh,
pp. 86-111

Churchman, C.W. (1971) The Design of Inquiring
Systems, Basic Books, New York

Commons, J.R. (1950) The Economics of Collective
Action, Macmillan, New York

Commons, J.R. (1924) Legal Foundations of Capitalism, University of Wisconsin Press, Madison,
Wisconsin

Dunn, E.S., Jr (1974) Information Processing and
Statistical Systems - Change and Reform,
Wiley, New York

Forsyth, F. (1980) interview reported in World
Press Review, 27, p. 61

Gale, S. (1972) 'Stochastic Stationarity and the
Analysis of Geographic Mobility', in
W.P. Adams and F.M. Helleiner (eds.),
International Geography 1972, University of
Toronto Press, Toronto, Vol.2, pp. 901-4

Gale, S. (1975) 'Boundaries, Tolerance Spaces,
and Criteria for Conflict Resolution',
Journal of Peace Science, 2, 95-115

Gale, S. (1976) 'A Resolution of the Regionalization
Problem and its Implications for Political
Geography and Social Justice', Geografiska
Annaler B, 58, 1-16

Gale, S. (1977a) 'A Prolegomenon to an
Interrogative Theory of Scientific Inquiry',
in H. Hiz (ed.), Questions, D. Reidel,
Dordrecht, pp. 319-45

Gale, S. (1977b) 'Ideological Man in a Non-ideo-
logical Society', Annals of the Association of
American Geographers, 67, 215-30

Gale, S. and Atkinson, M. (1977) 'Fuzzy Regions
and Social Justice', paper presented at the
meeting of the Operations Research Society of
America, Atlanta, Georgia

Stephen Gale

Gale, S. and Atkinson, M. (1979) 'On The Set
    Theoretic Foundations of the Regionalization
    Problem', in S. Gale and G. Olsson (eds.),
    Philosophy in Geography, D. Reidel, Dordrecht,
    pp. 65-107
Gale, S. and Moore, E.G. (eds.) (1975) The
    Manipulated City, Maaroufa Press, Chicago
Gale, S. and Olsson, G. (eds.) (1979) Philosophy in
    Geography, D. Reidel, Dordrecht
Georgescu-Roegen, N.(1971) The Entropy Law and the
    Economic Process, Harvard University Press,
    Cambridge, Massachusetts
Grigg, D. (1967) 'Regions, Models, and Classes',
    in R.J. Chorley and P. Haggett (eds.), Models
    in Geography, Methuen, London, pp. 461-509
Habermas, J. (1972) Knowledge and Human Interest,
    Beacon Press, Boston
Harvey, D. (1973) Social Justice and the City,
    Johns Hopkins University Press, Baltimore
Holland, S. (1976) Capital versus the Regions,
    Macmillan, London
Isard, W., Smith, T.E., Isard, P., Tung, T.H. and
    Dacey, M. (1969) General Theory: Social,
    Political, Economic and Regional, Massachusetts
    Institute of Technology Press, Cambridge,
    Massachusetts
Korner, S. (1966) Experience and Theory, Humanities
    Press, New York
Korner, S. (1976) Experience and Conduct, Cambridge
    University Press, Cambridge
Lear, J. (1978) 'Going Native', Daedalus, 107,175-88
March, J.G. and Simon, H.A. (1958) Organizations,
    Wiley, New York
Moore, E.G. (1979) 'Beyond the Census: Data Needs
    and Urban Policy Analysis' in S. Gale and
    G. Olsson (eds.), Philosophy in Geography,
    D. Reidel, Dordrecht, pp. 269-86
Moore, E.G. and Gale, S. (1973) 'Comments on Models
    of Occupancy Patterns and Neighborhood
    Change' in E.G. Moore (ed.), Models of
    Residential Location and Relocation in the
    City, Studies in Geography 20, Northwestern
    University, Evanston, Illinois, pp. 135-73
Nelson, R.R. (1981) 'Assessing Private Enterprise:
    An Exegesis of Tangled Doctrine', Bell Journal
    of Economics, 12, 93-111
Olsson, G. (1975), Birds in Egg, Department of
    Geography, University of Michigan, Ann Arbor,
    Michigan
Prescott, J.R.V. (1965) The Geography of Frontiers
    and Boundaries, Aldine, Chicago

Roweis, S.T. and Scott, A.J. (1978) 'The Urban
    Land Question', in K.R. Cox (ed.),
    Urbanization and Conflict in Market Societies,
    Maaroufa Press, Chicago, pp. 38-75
Suppes, P. (1962) 'Models of Data' in E. Nagel,
    P. Suppes and A. Tarski (eds.), Logic,
    Methodology and Philosophy of Science:
    Proceedings of the 1960 International Congress,
    Stanford University Press, Palo Alto,
    California, pp. 252-61
Toulmin, S. (1969) The Uses of Argument, Cambridge
    University Press, Cambridge
Toulmin, S. (1977) 'From Form to Function:
    Philosophy and History of Science in the 1950s
    and Now', Daedalus, 106, No. 3, 143-62

Chapter 3
INSTITUTIONS AND RURAL DEVELOPMENT

Gordon Clark

Studies of rural development have tended to adopt
one of three approaches based on either chronology,
location or land use.   Each of these approaches
can provide a useful structure for discussing rural
development, but equally each tends to mislead the
reader, albeit inadvertently.   The chronological
approach treats rural development in its historical
context, reporting and interpreting events.   This
has tended either to treat all events as unique or
to be excessively constrained by historicism in the
Popperian sense.   The second approach is locational,
looking at areas, usually states, and describing
each separately.   This approach highlights the
distinctive features of each country's experience,
but rarely succeeds in investigating the extent to
which motives and methods are similar in different
countries.   The land-use approach, which isolates
agriculture from forestry, recreation and other
activities, parallels the fragmented administrative
structure and the sources of data which national
governments employ for these land uses.   This
approach works well where decisions are taken, and
policies implemented, relatively independently for
each land use.   When this assumption fails for
either the private estate or the public domain
the land-use approach to rural development becomes
misleading.
     Any organising principle will tend to facilitate
the description of some aspects of the countryside
and obscure others which will be illuminated by the
use of a different approach.   The chronological,
locational and land-use approaches each contribute
to our understanding of rural areas but are ill-
suited to the comprehensive study of one aspect of
the countryside which has been assuming increasing
influence in developed countries.   This concerns

the effect of public and private institutions on
rural areas.   These institutions range from private
industrial and financial companies through public
corporations and quangos (in the UK, quasi-autonomous
non-governmental organisations) to the departments
of local, national and supra-national government.
Their effect is often mentioned by authors using the
three traditional approaches but the discussion is
usually conducted in a minor key and under the
assumption, often implicit, that each country's
institutions are historically unique (Edwards and
Rogers, 1974; Tarrant, 1974; Bowler, 1979; Gilg,
1978; Found, 1971).   These approaches are not well
suited to a systematic treatment of institutions
since they cannot answer questions about the
relative importance of different institutions, or
whether their effects in different countries are
similar, or if one can generalise as to institutions'
motives and methods for rural development.   The
institutional approach treats such questions as
central and so complements the other three
approaches.   The thesis of this chapter is that
there are insights to be gained from placing institutions at the centre of the analysis of rural
development in the developed countries.   The
chapter concentrates on the actions of governmental
institutions in the developed world and considers
private institutions only in so far as they impinge
on government activities or have elicited a response
from government.

Objectives of Government Institutions

If the current interest in the decision-making of
individuals were transferred to institutions, then
one would be concerned to elucidate their objectives
and motives.   However, several writers have been
at pains to point out the difficulty academics have
in this regard.   After many years teaching and
researching into British social policy, Brian Abel-
Smith entered the Civil Service at Deputy Secretary
level as a political advisor (Abel-Smith, 1980).
Seven years in this post forced him to revise many
of his ideas on policy formulation and how government worked.   Another former academic, Richard
Crossman, spent many years as a Member of Parliament
before he became a Minister and then a Secretary of
State (Crossman, 1975; 1976).   Despite his long
Parliamentary training, the realities of ministerial
office and Cabinet decision-making differed

markedly from his expectations. Consequently, the
fact that the present author is an academic outside
government and commerce must be borne in mind,
particularly since a government's stated objectives
may be very different from their real ones. The
safest initial assumption must be that institutional
motives and objectives may be even more varied,
complex and elusive than those of individuals.

It is unlikely that the actions of a democratic
government will be solely the product of one person
and traditionally government objectives have been
expressed in collective form. The work of
agricultural economists such as Whetham (1960, pp.
2-7) has tended to analyse government policy in
purely economic terms. She identified four objectives for agricultural policy: maximising the
national output of food, securing food supplies,
stabilising agricultural incomes and equalising
farmers' incomes. To her list should be added
improvement of the balance of payments, reducing
inflation and political advantage. One is struck
by the prominence of national and financial objectives in this list. These are often pursued even
to the extent of sacrificing progress made in
achieving more immediately agricultural objectives,
particularly at the level of the individual farmer.
Thus in the United Kingdom (UK) the system of price
support used by the European Economic Community
(EEC) to raise farmers' incomes was effectively
reversed between 1973 and 1979 by means of the
green currency mechanism. This lowered farmers'
incomes in order to reduce food prices and hence
lower the rate of inflation and wage settlements.
Thus a policy introduced to help rural areas was
subsequently used for the opposite effect with no
change in laws or institutions. The implementation
of a policy is as important as the form of the
policy. Similarly, a scheme for the enlargement of
farms is common to all the member states of the EEC,
but since the financial inducements have not been
raised in line with inflation the policy is
effectively under-used by those states such as the
UK which give only the statutory minimum grant, in
contrast to other states, Eire for example, which
give higher grants.

The list of government objectives is a long one
and this is inevitable given the desirability of
some progress towards each for whichever government
is in power. Most governments in the developed
countries are either a coalition of separate parties,
as is common in Scandinavia and the Low Countries,

or a coalition of different interests within a broad party as is normal in the UK, USA, France and, some would argue, the USSR.   Similarly, national Civil Services can be seen as permanent coalitions of departmental positions.   Policies with several objectives can be valuable in legitimising the government with many sections of the population and Civil Service by giving each interest group the prospect of some action of which it will approve. Poole (1970, p. 313) has noted how multiple objectives also arise from genuine debate over the true nature of the problem to be tackled.   Thus, low farm incomes may be ascribed to miserly subsidisation or the inefficiency of farmers or to excessive numbers of farmers.

Consequently a change of government leads to less alteration of policy than might be supposed and this is particularly true of agricultural policy in the developed countries.   The work of Rose (1980) has shown how, at least in the United Kingdom, political parties tend to carry out their electoral promises but that the difference between the parties' promises is not very great. Particularly in rural matters, most of the differences between the parties can be catered for by changes of emphasis in the implementation of existing measures.   The consensus between the parties is not complete but it is at least as substantial on rural topics as for any other issue in Britain and America.   This consensus is not always as strong in Western Europe, and cases of sharp reversals of rural policy can be found, for example, between the Weimar Republic and the Nazi Government in inter-war Germany (Huggett, 1975, pp. 141-3).

This tendency to stability can lead to a policy being used to achieve a different objective from the one for which it was introduced.   The forestry policy of the United Kingdom provides a clear example of this.   As late as the 1950s, the primary objective was the provision of a secure supply of timber from domestic sources in the event of war.   By the 1960s, forestry had become important principally as a source of full-time employment in rural areas marked by out-migration and declining employment in agriculture.   The policy was the same but the original objective had been superseded to the extent of almost ceasing to have any force. Similarly in the United States the support of farmers' incomes was born of the poverty among small-scale farmers particularly in the late 1920s. Today the same policy, albeit achieved by a

different mechanism, is defended as much for its
beneficial effect on the balance of payments as for
its effect on farmers' incomes, which is in fact
regressive as it benefits the smallest-scale
farmers the least (OECD, 1974, p. 30).

Initiatives for rural development may also be
introduced for reasons only indirectly connected
with rural welfare. In the United States, the food
stamp programme was partly a device to bolster
demand for food but more importantly it was a social
welfare measure for the urban needy, moulded by a
reluctance to give them money which they might not
spend on food. Farmers and the lobbies in Congress
for agricultural commodities obviously supported
this measure since the food stamp programme was to
their immediate advantage.

Governments act slowly and policies are rarely
implemented quickly. Indeed, there may be such a
long gap between recognition of the need for action
and measures being implemented, that conditions may
have so changed that the policy is out of date by
the time it is introduced. One might interpret
the history of rural housing policy in Britain in
this way. In 1900 rural housing conditions for
many people were very poor but it was not until
after 1945 that a programme of council house construction in rural areas began on a scale commensurate with the need. Yet by then the problem had
been greatly reduced in scale by out-migration, city
dwellers renovating property and by self-help at the
expense of investment in farming by householders
and tenants. Council house building had not become
irrelevant but the lag between the identification of
the problem and action being taken was so long that
the need for the policy had been greatly reduced.

It is not uncommon for a well-tried policy to
continue to be used even in circumstances where a
fresh look at the situation would suggest different
action. In the EEC the perennial over-production
of milk, itself a product of subsidies to farmers,
led to two initiatives. The first subsidised the
conversion of liquid milk into skimmed milk powder
which was used, with more subsidies, as a feedstuff
for young dairy calves which in their turn exacer-
bated the over-production of milk. The second
initiative subsidised farmers to convert from dairy
to beef cattle. There is now a surplus of beef as
well as dairy products.

In the United States in the 1950s, there was
considerable disquiet over the cost of storing
surplus grain. A system of acreage quotas was

79

introduced to help reduce the surplus by paying
farmers either not to produce a given crop on an
agreed area, or not to produce any crops on that
land.   The former course had the unplanned result
of transferring the surplus from wheat to sorghum
which many farmers grew instead.   The alternative
of banning all production under the Conservation
Program of the Agricultural Act of 1956 would have
worked well if yields from the remaining area had
stayed constant.   Instead they rose as the payments
to stop production could be spent on additional
fertiliser for the remaining land.   Surpluses are
no longer so serious a problem as in the 1950s, not
so much because the acreage 'set-asides' have worked
as intended, but rather because the export demand
has risen sufficiently to remove the surplus - a
trend encouraged for its balance of payments and
foreign policy benefits.   Where payments for
acreage 'set-asides' continue to be made, despite
being only half as effective as planned, they can be
seen simply as income supplements to farmers (Mann
and Dickinson, 1980).   The policy of 'give them a
grant' is sometimes the standard response by governments even when a lack of money is not the
difficulty, as with the grants to improve the poor
management standards on small farms (Robson, 1971).
It is not clear whether this is because grants are
popular even when ineffective, or easier to devise
and administer than better-directed measures.
        One can also detect some distinction between
the objectives of state policies in several
countries before and after the Great Depression,
most notably in Britain and America.   Before the
late twenties, state policy for rural areas was
laissez-faire, particularly in America after
initial settlement had been achieved under the Homestead Act.   Government action was generally confined to censuses, farm education and research
which were all broadly favourable to rural development and economic growth.   Other policies were concerned with heading off local opposition or civil
unrest focused on a particular grievance such as
the laws of land tenure in Britain.   The Irish
Land Acts from 1870, the Crofters Holdings
(Scotland) Act of 1886 and the legislation on
tenancy in England and Wales after 1875 all shifted
the balance of short-run advantage against landowners in response to political pressure.   The
Scottish and Irish measures were more radical than
those in England and Wales.   It would appear that
on the rare occasions far-reaching initiatives are

taken, they are introduced in the more distant parts
of the state so as to minimise the risk to the core
of the state if they prove dangerously radical.
The history of regional development organisations in
Britain, Italy and Norway suggests the same tendency
to experimentation at the periphery of the state
rather than in its core.   In the case of land
tenure, the intention was to secure the government
in power despite strong opposition and such policies
can be seen as an example of the state as arbiter
(Dear and Clark, 1978).

Policies for rural development since the Great
Depression have tended to be more explicitly inter-
ventionist, partially supplanting the market
mechanism in determining the worth of goods and the
incomes of different groups.   This was not so much
in response to specifically articulated threats to
the government's continuation, rather it was an
attempt to improve the growth and stability of the
whole economy.   To a limited extent this was a mani-
festation of the acceptance of Keynesian economic
principles although rural development has always
been rather peripheral to policies of demand management.

Although the intervention was concerned to
raise national economic performance, it did not
always benefit all sectors of the economy nor all
groups in society.   In some circumstances, governments have reduced living standards in the countryside in order to meet a national economic objective.
In the 1930s, for example, the Soviet Government
kept farm incomes low and sold food at a profit so
as to generate surplus for investment in industrial
expansion.   Similarly, Britain saw cheap food and
low farm income as the best course during her
industrialisation after 1846.   The opposite policy,
the subsidy of Soviet farmers' incomes from general
taxation, has been used in the 1970s as an incentive
to raise the volume of food produced to help the
balance of payments as the demand for food,
particularly meat, rose in line with increasing
urban living standards.   The USSR's version of the
price mechanism has been used in ways analogous to
post-war British and American policies to aid
national economic development in the manner best
suited for the period in question, even when rural
producers suffered to a greater or lesser extent
because of this.

In addition, where the promotion of rural
development has raised rural living standards, this
has proved to favour disproportionately certain

groups of rural capitalists.   Many farmers are in
the unusual position of being capitalists in the
sense of being owners of land, employers of labour
and creators of surplus value and yet they are
exploited themselves.   The small-scale farmer
receives least subsidy and the policy of under-
remunerating farmers for increases in their production costs favours the large-scale and more
efficient farmer at the expense of the small-scale
farmer.   The widespread policy of reducing the
number of small farms requires the existence of
sectional inequalities of income.   If the small-
scale and less efficient farmers were not kept poor
they would not have any incentive to leave the
industry.   Farm policy may be counter to the
interests of some rural capitalists if this is in
the broader interest of national economic growth and
efficiency or the welfare of the agricultural sector
as a whole or of its more influential members.   The
contrast between the periods of laissez-faire and
intervention should be interpreted therefore as a
change of emphasis rather than a revolution.   In
the later period there is less concern with fighting
serious grievances and avoiding impending rebellion
and more with aiding farming, but only in so far as
this is consistent with national economic or political objectives.
    The argument that agricultural policy may be
inimical to the interests of some capitalists
applies with particular force when development over
the long term is considered because of a tendency
for policy to be short-sighted.   The long-term
costs and benefits from a policy tend to be given
very little weight in decision-making whereas outcomes in the near future are given disproportionate
consideration.   This may be because the decision-
maker will not be around in twenty years' time to
reap the fruits or whirlwinds of his decisions.
There is also a tendency to argue that since 'a week
is a long time in politics', the next decade must be
utterly unpredictable and far too uncertain to
bother about.   Public policy may be based on such
short time-horizons that there may be occasions when
it can best be characterised as rapid reaction to a
stream of crises with decisions almost taking themselves (Abel-Smith, 1980).   To attribute clear goals
to public decision-making may sometimes be to make
the unwarranted assumption of deliberation and
rationality.   Decision-making implies a choice
between rival policies yet if policy is for the
short term or is a hasty reaction to a crisis thrust

upon a government, then there may be only one
feasible policy to follow.   In these circumstances
governmental decision-making is the process of <u>not</u>
making genuine decisions since there are no choices.
    This myopic quality in public policy is illus-
trated very clearly by the cost-benefit analysis
conducted by the British Treasury as part of a
review of forestry policy (Price, 1973; 1976).   The
study, which attempted to measure the current and
future costs and benefits of forestry, incorporated
a high discount rate on future benefits which
effectively placed a very low value on timber production in the distant future when trees planted now
would be harvested.   The choice of discount rate in
this study not only created an economic case against
public funds being used for forestry, it also
typified the overriding concern for the near future
in public policy.
    It is not only temporal horizons which are fore-
shortened in rural policy since spatial horizons are
also restricted, particularly by national boundaries.
The EEC, for example, is much concerned to promote
stability of European farmers' incomes and one means
of achieving this at a time of glut is to subsidise
the export of excess food produced in the Community.
For countries outside the EEC which depend on
exporting food each year, these irregular supplies
of subsidised food create considerable economic
instability.   For countries wishing to promote
their domestic agriculture, these irregular cheap
imports may hamper steady growth, although the
effects of cheap food imports to these countries
may not all be disadvantageous (Schuh, 1979).   Thus
the EEC achieves its stability by helping to create
instability in some other countries.   Wallerstein's
review (1981) of the US food aid programme also
judged that national self-interest rather than
altruism was the principal inspiration for much,
though not all, of that policy.   Rural policy may
therefore be constructed within very narrow temporal
and spatial horizons in the name of 'national or
supra-national interest'.

Institutions' Methods

There are relatively few methods by which governmental institutions have attempted to achieve their
aims in rural areas, the principal ones being landownership, subsidies, development control, provision
of infrastructure, fiscal measures and, finally,

ad hoc agencies.

Landownership is the most direct method for institutions to achieve their objectives but also the most expensive both in terms of capital - land is expensive particularly in Western Europe - and also in recurrent costs for administration and investment.

The ownership of land by the state has a long history. The Northfield Committee estimated that public bodies owned about eight per cent of farm-land in Britain while in Scotland the proportion of land publicly owned has increased from under one per cent in 1872 to fourteen per cent in the late 1970s (Northfield, 1979, p. 50; Clark, 1981). Public landownership has been used to fulfil a variety of objectives including the provision of small farms in Italy, Norway and the United Kingdom, forestry, defence, land improvement, nature con-servation and the protection or containment of native peoples.

In the United States, Federal landownership is considerable, accounting for one-third of the entire country. In the western states the Federal estate makes up over half the land area - in Nevada it is eighty-five per cent and in Alaska ninety-six per cent (Jackson, 1981, p. 48). In these areas the policies of the Bureau of Land Management, the Forest Service, the Department of Defense, the Fish and Wildlife Service and the National Park Service are the principal, or even the sole, determinants of land-use policies over vast tracts of the country. State bodies also have an effect on land ownership though a lesser one - in California seventeen state agencies or institutions own land in the state (Jackson, 1981, p. 63).

Farmers tend to object to the involvement of the state in the land market yet have a contradictory need for state involvement in food markets. In the developed world outside Eastern Europe, land nationalisation on a large scale has not been attempted and even in Eastern Europe, rural Poland and Yugoslavia are mostly privately owned. Experience shows that a ratchet principle operates with public landownership in that it is harder to sell land than to buy it even when the need for, say, state smallholdings is long past. The current proposal to sell some public forestry land in Britain is therefore an interesting experiment in partial denationalisation. The state can also become involved in the land market in a temporary way where improvements in farm structure are needed.

In France, the regional organisations called SAFERs
control the sale of land so as to create larger
farms while in Britain a brief experiment along
these lines was carried out by the North Pennines
Rural Development Board from 1968 to 1970.

The state can also become involved in the land
owned by private institutions where this reaches
such a level that the public interest seems to
warrant regulation of that ownership to prevent a
monopoly developing or the abuse of power.

Private institutions such as insurance
companies are increasingly buying land rather than
lending to landowners and this has aroused some
discontent in the United Kingdom and the USA. The
Northfield Committee estimated that in Britain
financial institutions owned just over one per cent
of farmland in 1977 (1979, p. 60). Their motives
for buying up to ten per cent of the farm land for
sale annually since 1974 have been the growth in
farm rents and rapidly rising land values based on
Mark Twain's dictum about land - 'They ain't making
it any more.' Farmers allege that the rise in
land prices is partly the result of the institutions'
presence in the very small and imperfect land
market. Private farmers are faced with a dilemma
since they approve of profit making and good
husbandry and yet tend to fear and object to institutional landowners particularly when they invest
heavily and make large profits. In the United
States, a parallel debate concerns the growth of
agribusiness which is the vertical linkage of
industrial and food processing companies and large
corporate farming organisations (Smith, 1980).
The amount of land they control is far smaller than
the proportion of production for which they are
responsible. The debate contains contradictions
such as approval for private individual ownership
of land but not for private corporate ownership.
There is also a strong but ill-argued belief in
the desirability of small-scale and family farms
which are claimed to be threatened by agribusiness
(Mann and Dickinson,1980; Burbach and Flynn, 1980).

The second method of influencing rural areas is
by subsidies. These can be given for farm products
by means of deficiency payments - the cost of these
is met by the taxpayer so the poor, who pay little
tax, benefit most - or by intervention buying and
import taxes to raise food prices - the low income
earners are affected most severely since this method
raises food prices - or by non-recourse loans.
These loans are repayable if the market price is

high and are kept as a subsidy if the market price
is low.   The latter method along with storage and
subsidised exports was the mainstay of the commodity
support schemes in the USA for many years while the
deficiency payments system has also been used at
various times in America and in the UK between 1947
and 1973.   The EEC uses it for a few products but
mostly relies on import taxes and intervention
buying.   Subsidies are also given by the EEC and
national governments for forestry, investment in
buildings, land improvement and farm enlargement.
The first difficulty with subsidies is that the
recipients may act in the same way with the subsidy
as without it.   The lime and fertiliser subsidies
in Britain in the 1960s became free gifts to farmers
since the rate of subsidy was only weakly related to
the volume of lime and fertiliser used.   Similarly
the farm amalgamation schemes in the UK probably had
relatively little effect in raising the number of
amalgamations.   It is very difficult to measure the
effectiveness of the subsidy since the counterfactual
is largely conjecture.
     The second difficulty is that many subsidies
are regressive in that they usually give most to the
largest-scale producer.   The largest five per cent
of all farms in the USA received 42.4 per cent of
the subsidies in 1969 and the smallest twenty per
cent of farms received only 1.1 per cent.   However,
the small sums received by the smaller producers
probably have a high marginal utility.   The justification for subsidy schemes in most countries
includes supporting the incomes of poorer farmers
and helping investment by less efficient farmers,
yet most aid goes to those least eligible on those
grounds, although it could be argued that large-
scale farmers are, like British Leyland and Chrysler,
too important not to subsidise.   It has often been
proposed that the level of subsidy should be reduced
by setting it with reference to the needs of
efficient farmers.   The task of supporting the
incomes of small-scale and poor farmers would then
be transferred to the social security system.   This
would avoid such distortions in food production as
excessive dairy production which are caused by
aiding both large and small-scale farmers.   However,
only Sweden has systematically sought to reduce its
self-sufficiency in food by reducing farmers'
incomes (OECD, 1972, p. 225).
     In some countries income support for farmers
also has a role in regional policy.   Norway and
Switzerland give specially generous subsidies for

farmers in remote or mountainous areas while in
1975 the United Kingdom persuaded the EEC to adopt a
version of its scheme of hill subsidies which was
designed to recompense hill farmers for the limited
benefit they obtained from other subsidies.   These
special subsidies are believed to reduce rural
depopulation, maintain communities and even preserve
the traditional upland landscape as neat hay meadows
and pasture rather than untidy scrub land.

The third method of influencing rural areas is
by development control.   This has evolved in Britain
since the Town and Country Planning Act of 1947 and
is a system whereby changes in land use in the
countryside require permission which, in theory,
allows for public control of change at the margin.
As used in Britain, the degree of control over rural
areas was much less than in the cities since many
changes of land use involving agriculture, forestry
and defence were in effect approved automatically.
This was partly due to pressure from the very well-
organised farming lobby in Britain and partly
resulted from a failure in 1947 to foresee the
recent pressures on the countryside.   This lack of
control over rural Britain is being rectified to a
limited extent in the National Parks and where
agricultural grants are given.   Even where control
has been exercised, its effects have been questioned
by several authors.   Blacksell and Gilg (1977)
found that control was sometimes more lax in areas
where the landscape was to be protected than in un-
protected areas although Anderson (1981) came to
the opposite conclusion in her study of Sussex.

In Norway and Denmark control of land use was
achieved by zoning some rural areas for particular
uses.   Some parts of Denmark were zoned for second
homes which were banned from other areas while in
Norway protection for the wild beauty and ecology
of the highest mountain areas was given by zoning
most development away from these areas.   Zoning is
also a favoured method in the United States where
states and counties have very considerable autonomy
in rural planning to the extent that some practice
almost none, while others have schemes which are
unenforceable and a limited number of states like
Vermont, Hawaii and coastal California have
elaborate and strictly enforced schemes for regulating land use (Healy, 1976; Geisler, 1980).

Two fundamental questions have to be asked
about all systems of zoning and development control
since they will affect land use and hence land
values.   First, should private landowners reap the

87

benefits or suffer the losses created by decisions
on land use taken in the public interest by local or
national government?   In the British system, the
horns of this compensation-betterment dilemma have
never been firmly grasped although for short periods
windfall capital gains have been heavily taxed.
The British system, where compensation is not paid
for losses to landowners contingent on planning
decisions, can be less restrained in its zoning
than can the planning systems in Norway and Denmark
where compensation must be paid.   Clearly the
artificial restriction of the supply of land with
permission for a particular form of development is
likely to inflate land prices and benefit those
fortunate enough to own the land (Hall, 1974, pp.
403-7).

Second, who is to benefit from land use control?
It is not clear who benefits from the confinement of
new houses in Britain to the existing settlements
rather than the open countryside.   In the United
States, the enforcement of a quite large minimum
size of building plot benefits the neighbouring
house owners by keeping up house values and it
benefits the municipality by keeping out those who
would demand public expenditure on welfare services
and contribute little to local revenue.   The central and unresolved issue with land-use planning is
the distributional question - the effect of controlling land use and development rights on land prices
and the incomes and wealth of different groups in
society.   This is still a major issue in the United
States whereas in Britain debate on land-use
planning centres more on the means of achieving it
than its distributional consequences.   In neither
country is it clear what is the precise meaning of
the 'national interest' so often invoked as justification for landscape preservation, for example
(Department of the Environment, 1974).

The fourth method of government influence is
through the provision of infrastructure.   This has
assumed increasing importance as the general
standard of services and facilities has risen.   In
every country the provision of services is more
expensive per head in rural areas than in the towns
because of the low density of rural populations even
though the total cost of providing the facilities
may not be very great.   Consequently the diffusion
of public services tends to be hierarchical down the
urban system with the rural areas furthest from the
cities receiving the service last.   In England and
Wales only eight per cent of farms in 1938 had a

mains electricity supply while two-thirds of
parishes had no mains sewage scheme in 1942 and
seventy per cent had no piped water supply although
both were standard, even if inadequate, in towns and
cities (Ministry of Works and Planning, 1942, p. 19).
    In some countries such social provision has
been viewed as a major element in regional policy
to stop rural depopulation. However even when as
wealthy a country as Switzerland takes this view, it
is scarcely possible to afford a school and hospital
in each hamlet. The difficulty lies in finding a
rational basis for allocating such facilities as can
be afforded.
    The commonest solution is the designation of
some settlements for a wide range of facilities
while others receive little or no new public investment. This is known in Switzerland as dispersed
concentration and in Britain as key settlement
policy and has been incorporated in Soviet rural
settlement planning (Grafton, 1980). It has the
merits of financial expediency and inevitability
but its operation raises serious and difficult
questions about how the designated settlements are
selected. There is no theoretical guidance on how
many key settlements an area should have nor on how
large a key settlement should be. Ad hoc solutions
are used with political influence playing a part
too (Cloke, 1979). There is also a belief that
concentrating public housing and other facilities
will promote additional economic growth through an
enhanced multiplier effect but this appeal to
growth pole theory has been criticised for having a
weak theoretical basis (Cloke, 1979). In so far
as key settlement policy provides a superficially
plausible rationale for providing fewer services
and so for not spending money, it is likely to be
as welcomed by local councils dominated by rate-
payers (Newby, 1979, p. 194) as by a centralised
command economy such as that of the USSR (Pallot,
1979).
    The concentration of facilities will work
equitably if public transport is cheap and at convenient times of the day as in Switzerland or if
most of the rural population has access to a car
without undue financial strain as in much of North
America. The British case appears to be one where
there is insufficient money either to disperse
facilities or to provide a cheap, convenient form
of public transport. Rural people may have to buy
a car although the lower income groups really cannot
afford one. The rate of car ownership in rural

Britain is ten to twenty per cent higher than in urban areas, although average earnings are lower in rural areas which, at least in the case of rural Wales, have similar income distributions to the United Kingdom as a whole (Thomas and Winyard, 1979). Once people have a car, enticing them back on to public transport is very difficult. Alternatively, they eventually migrate to the towns where their lack of a car is a less serious handicap. There is therefore the prospect of British rural areas being populated by the wealthier who can afford personal mobility and by those who have to live and work in the countryside and who gain access to a car whatever the cost. The policies of private individuals and companies play some part in this through the closure of village shops, for example, but the major role is taken by public bodies who have to decide the terms on which they allocate facilities and the transport to them.

On the other hand, there is a long history in most countries of hidden subsidies to rural areas through the pricing policy of public and private bodies. Water, electricity and postal services are usually provided for rural areas at substantially the same cost to the consumer as in the cities. The service may not be as good with fewer postal deliveries or a manual telephone exchange but rural users are in effect subsidised by urban consumers. Subsidy from uniform pricing also applies to private companies supplying food and other goods at a single national price. This acts as a partial counterweight to the higher shop prices in rural areas engendered by lack of competition between shops and their low volume sales.

The fifth method of influencing rural areas is through the taxation system. The differential effects of taxation on rural and urban areas is an under-researched area but some consequences are known. In Eire and West Germany, few farmers pay income tax even though they would have been liable to tax if the income had been earned in another industry. In Britain, the operation of Schedule D of Income Tax has a similar though less marked effect in reducing liability to tax. The opportunities farmers have for tax evasion also have to be borne in mind. In Britain, farmers have historically received some concessions with respect to inheritance taxes and the same is true of France, Italy, Japan and Eire. British farmers pay no tax on the diesel fuel they use and recently Dutch horticulturalists and French fishermen have also been able to buy cheap fuel.   For many years, British private landowners have received very favourable treatment under Income Tax and Capital Gains Tax if they afforested their land.

Conversely rural people, with greater distances to cover, can suffer unduly from the national application of extra taxes on fuel so as to raise revenue or cut consumption.   Tax concessions have usually only been given to rural farmers whereas higher taxes tend to affect the whole rural population.

The final method of influencing rural areas is through ad hoc agencies established by government to carry out specified tasks.   Often these agencies deliberately transcend local government boundaries and many are concerned with economic development, particularly industry and tourism.   The role of the Tennessee Valley Authority soon came to encompass such a development role while the North Norway Fund and the Cassa per il Mezzogiorno in Italy are other examples.   In Scotland, the Congested Districts Board (1897) operated along similar lines while the North of Scotland Hydro-Electric Board (1943) was also given an industrial development remit.   Since 1965 the Highlands and Islands Development Board has taken over these functions while in England and Wales the Development Commission (1909) and Welsh Development Agency (1975) have followed the path of publicity, industrial promotion, sales missions, factory provision, advice to small businessmen and loans to potential entrepreneurs.

These agencies find tourism quite easy to promote but the employment generated tends to be seasonal, especially in the remoter areas.   The amount of tourist expenditure which stays in the local economy - the income multiplier - is relatively small particularly for hotel-based tourism.   Once the decline of farm employment has been recognised, industrial development becomes the ideal source of full-time, year-round jobs in the face of a contracting primary sector.   It has always been very difficult to establish industry in the countryside once existing craft industries have been swept away by cheaper products from large automated factories.   Consequently, it is arguable that the countryside is now more predominantly agricultural than at any time in history.   Gaskin (1971) concluded that difficulties in assembling a labour force with the required skills were the principal obstacle to the industrialisation of the major Scottish islands rather than the higher transport costs which were conventionally blamed.   Where

rural industry has been established, it is usually
part of the ever-expanding sphere of influence of
major cities or along motorways, both trends being
most marked in the USA.

Some industrial development has been of a
character rather unsympathetic to rural communities.
'Cathedrals in the desert' is the rather cynical
name for the large steel and car plants established
in southern Italy since 1945. The labour force may
have to be imported if the required skills are not
available locally or if the local labour market is
too small. As with oil developments in Alaska and
northern Scotland, this can cause social and
cultural tensions, inflation and labour poaching
from established local firms (MacKay and Mackay,
1975, pp. 111-37). Defence establishments are
sometimes located in rural areas for strategic or
security reasons and can generate both permanent
jobs for local people and tensions and opposition to
their presence. The ideal source of rural employment would be locally owned and would employ a small
number of well-paid men and women all year making a
product for which demand was rising faster than
labour productivity. In the absence of this
counsel of perfection, governments still seem to take
the attitude that rural areas should take what they
can get and let migration and social security deal
with the unemployed.

Each of these six methods of controlling rural
development has been used by most governments at one
time. Most have ad hoc agencies for rural planning
or development and tax systems favourable to farmers
though rarely to other rural people. Every country
protects its agriculture, especially America and
Japan, even though these two governments would
blanch at the prospect of similar treatment for
industry. Without wishing to argue that countries
are indistinguishable from each other in the treatment of their rural areas, nonetheless it is remarkable how many have pursued at some time in their
history similar policies by similar methods. This
is partly due to the common path for the evolution
of rural areas in so many countries with rural depopulation, declining farm employment and the
growing need for expensive infrastructure to
equalise rural and urban standards of living. Also
there are only a limited number of tactics for
meeting these changes which do not involve the
dangers of rebellion or the uncertainties of a
radical transformation of the rural economy. Most
individual countries show as much variation in the

history of their government's treatment of rural areas as there is variation between countries at the same point in time.

Features of Institutional Action

The theme of the similarity of institutional action in different countries can be developed further by highlighting other aspects which transcend national boundaries and land uses. So far, the discussion has been cast in terms of the intentions and methods of institutions with respect to a single problem. In reality, all governments are simultaneously operating policies for various problems in each rural area and the interactions among policies are similar in many countries.

Whereas Ministries and land uses may exist in neat compartments, the rural economy and society do not. Events are interrelated in the real world and whether from departmental rivalry, inadvertent lack of liaison or a failure to appreciate this interdependence, numerous examples can be found of the achievement of one policy's goals making it harder to achieve progress towards another goal. Sometimes the conflict is almost insoluble as in national parks where recreational planning is supposed to facilitate solitude for the masses in the once quiet countryside. At other times, the conflict arises from a failure to think through the juxtaposition of policies. Thus the priorities for forestry and landscape preservation in the treeless uplands of British National Parks have not been established, so pointless conflict occurs with every new tree planted.

Similarly, most governments give subsidies and tax relief to farmers. The effect of these is to raise farmers' incomes and so raise the amount they are willing to bid for land. In effect, the subsidies have been capitalised in higher land prices. Simultaneously, most governments have schemes to help farms expand but this structural policy is generally hindered by the higher land prices. This is particularly so for those farmers receiving least subsidy who have small cash reserves, small farms and so a low borrowing potential. Similarly, raising agricultural land prices makes the state forestry policy more expensive. Finally, as land prices rise more farmers will be liable to capital taxes despite the concessions they obtain and the tax may have to be paid by selling a part of the

Gordon Clark

farm or by reducing investment, neither of which are among the intended effects of agricultural policy (Evans, 1969). The extent of this division of farms will depend on the relative levels of land prices, farmers' incomes and tax rates in each country (Northfield Committee, 1979, pp. 71-90). Every tax concession raises the price of land as outsiders with high marginal tax rates struggle to get into farming to enjoy the lower taxes.

This interaction between policies within a situation of administrative fragmentation and departmental rivalry can be detected even in cases where a determined effort has been made to promote co-ordination. In the United Kingdom, key settlement policy was used by several counties as a means of justifying a concentrated distribution of public expenditure in rural areas (Cloke, 1979). The inability of the policy to work as intended has been partly a reflection of the extraordinary difficulty governments have in co-ordinating the investment programmes of their own agencies, let alone private bodies as well.

A second common feature of institutional action in rural areas is how regressive it can be. The British Government noted in 1965 that the small-scale farmer had the following stark prospects (Her Majesty's Stationery Office, 1965, p. 1):

> The Government believe that one of the more important problems facing agriculture today is that of the small farmer trying to earn a livelihood from insufficient land. As time passes, his difficulties will increase. Many of them [small farmers] may be able to increase the size of their business... But, looking ahead, many others, however hard they work and however well they manage their businesses, just cannot hope to get a decent living from their farms at prices which the tax payer and the consumer could afford.

The small-scale farmer gets least subsidy (OECD, 1974), he expands least frequently (Clark, 1979), he receives fewest grants for investment (Bowler, 1979) and generally is an embarrassment to government and a boon to large-scale farmers as a source of both casual labour and extra land so that they can expand their farms in manageable steps. When governments set price levels which will not distort production, they do so with reference to farmers of average scale and efficiency. This over-

94

recompenses the large-scale farmers for inflation while under-recompensing the small producer. Even taxation which is usually intended to be progressive may be regressive in practice. Thus 87 per cent of the lowest income American farmers were taxed on their profits in 1966 whereas only 39 per cent of the highest income farmers paid tax since they could offset their much larger profits against depreciation, for example (US Chamber of Commerce, 1974). It is rare indeed for small-scale farmers to be both the intended and actual beneficiaries of public policy for rural development.

The regressive effect of rural policy can also be seen where the full cost of rural travel has to be met by the traveller since this falls most heavily on the lower income groups. In the field of housing, the imposition of a quite high minimum size of building plot in many American counties militates against low-cost housing, as does the requirement for high standards of construction and design of houses in British National Parks. Also, the severe restriction of the number of new houses to be built in the Lake District National Park for aesthetic reasons will probably have a markedly inflationary effect on house prices and lower income groups will gain least from the policy. Hall has noted the inflationary and regressive effect of British land use controls since 1947 (Hall, 1974, pp. 403-7). The regressive effect of these measures is arguably inadvertent rather than deliberate, and this is an area which needs much more empirical investigation.

Many conflicts of interest exist in rural areas; in farming there are those between large- and small-scale farmers, livestock and cereal farmers, and upland and lowland farmers. The remarkable feature of farming in many countries is how united it is in its lobbying and how coherent in its self-perception as a single industry of farmers and workers despite these conflicts of interest (Newby, 1979, pp. 255-62). The main exception is the USA where the farm lobby mirrors the rest of the political system in its diversity (Wilson, 1978). If agriculture is still a community of interest, it is so because it feels itself trapped by real and imagined threats to the whole industry from politicians, townspeople and financial and industrial companies. Pressure groups such as COPA in Brussels and the NFU in Britain are models of integration. The few instances of break-away groups like the Farmers' Union of Wales and the American Agricultural Movement have had clear foci for their discontent but they have rarely succeeded in establishing and maintaining a power base.

Farmers are individualists but not to the extent of failing to realise that unity is strength. Only unity can protect individuality. Similarly, it is not in the interest of any government to legitimise such dissident groups. It is useful to negotiate with a well-known, predictable group of people who 'represent farming' since agreement is easier with one group than with several holding divergent views (Wilson, 1975). Such agreement is important despite the fact that policy can be uni-laterally enforced by government, since agreement signifies support from the moderate farmers' union against the minority of dissidents and it legitimises and conceals the power of government.

Where the differential between rich and poor people and regions has become so large that it reaches the threshold of tolerability, government will have to take palliative action to reduce the differential. The threshold of tolerable differ-entials seems to vary between countries, being higher in France than Sweden, for example. Normally the palliative is left to the tax and social security systems and subsidies to local government (e.g. Rate Support Grant) rather than being incorporated into specifically rural policies.

A third feature of rural policy is highlighted by the debate about the extent to which it is influenced by the Civil Service of the country con-cerned rather than by the politicians. This controversial topic is extremely difficult to research since self-effacement would be in the interests of civil servants if they had power and wished to keep it, while self-aggrandisement is not unknown among politicians. In the British context, Crossman (1975; 1976) characterised the Civil Service as an organisation of great influence by virtue of its continuity of personnel whereas politicians came and went. The Civil Service also has superior access to information and it scruti-nises many of the people and policy options to which the politician is exposed. Such a continuing influence on any government can be seen as acting for the status quo and against rapid or radical change in policy. Departmental positions need to be defended and practical difficulties in implementing any new policy will be pointed out. Similarly the dominance of economists among the professional and academic advisors to government may have had the effect of narrowing the perceived problems of rural areas to considerations of efficiency and equity measured in monetary terms. There is certainly a presumption in favour of rural development in a conventional economic sense in every country. Consideration of unquantifiable costs and of social and environmental matters are accordingly given less weight. Few countries have dared to be radically different in rural policy - the community co-operatives in Eire and Scotland are minor examples while the various forms of collective agriculture in Israel and Eastern Europe are more adventurous and better known. For whatever reason, rural policies tend to be characterised by continuity rather than change.

Whatever policies have been adopted for rural areas, the countryside in most countries has become less distinctive and self-contained. Dialects and minority languages retreat under pressure from the mass media and school education, rural crafts are supplanted by factory production or caricatured for tourists. The links between farming and industry are strengthened by the increasing dependence of farmers on tractors and chemicals instead of horses and local lime. Conversely, industry is taking a firmer control of farming by entering into direct contracts with farmers to produce agreed quantities of food for a fixed price. This removes much of the risk from farming engendered by volatile markets, but it carries with it a loss of independence as the farmer becomes like a factory worker - a skilled operative carrying out prescribed methods of production for known levels of remuneration. In some cases the involvement of major food companies in production has gone further, most notably in the American beef-feeding lots, as companies have taken to farming directly. One of the developments in Soviet agriculture in the 1970s was a similar trend to link more closely farm and factory into 'agro-industrial complexes'. In parallel to this, farmers have been combining into producers' co-operatives which move downstream into food processing and marketing. This is most successful when the co-operative imposes strict discipline on its members and employs professional staff to manage the processing and marketing operations. In short, a successful co-operative is like a company with the farmers as shareholders, although not necessarily the majority shareholders (Berger, 1971). Either way, industrial methods, values and disciplines come to be imposed on, or accepted by, farmers for the common good where this

Gordon Clark

is defined in a conventional economic framework.

Yet the increasing interdependence of agriculture with industry and state intervention coexists with an enduring stereotype of rurality which came to be expressed in the socio-geographical dichotomy of gemeinschaft and gesellschaft. Conventionally, rural areas are seen as archaic, simple, picturesque and quite separate from cities in their work, attitudes and values. Pahl's 'village in the mind' is the setting for the sturdy family farm - an enduring image which has had an influence on public policy in the USA, Italy and Scandinavia. That the image is no longer valid is less important than that it fulfils a common need for an Arcadia somewhere. The reality is that the countryside is neither backwater nor paradise. It has become an integral part of national economies, not indistinguishable from the cities to be sure, but bound up inextricably with them and no longer to be understood in isolation from them. The rural-urban dichotomy is of only residual utility and the actions of public and private institutions have contributed to this in no small measure.

Problems and Prospects

The institutional approach to the study of rural development has much to offer in improving our understanding of the relative power of individuals, the Executive, the Legislature and the Cabinet. It can also be the focus for comparisons between countries. It needs to go beyond the existing studies of specific policy initiatives,yet, to do this convincingly, there is a need for more inside information on how government works than most geographers now possess. The approach needs to generate high-level generalisations on the motivations for, and effects of, government action. Such models already exist and Dear and Clark (1978) describe several. While one can see traces of some of these in specific instances none really rings true in the rural context. They all have the caricatural quality of the economic-man model and are not very enlightening although straw men can have the value of stimulating critical thought.

The institutional approach to rural geography contains the twin dangers of anthropomorphising the state and relegating individual behaviour to the rank of trivial exceptions to general rules. The task as always in geography is to transcend

different scales of study to relate the actions of
individuals to each other, of governments to other
governments and of individuals to governments.
This is not easy since each scale of study has its
own sources of data, its literature and methodology,
its own concepts and styles of explanation.   The
institutional approach is not an end in itself, just
a useful means to a more rounded and stimulating
rural geography.

References

Abel-Smith, B. (1980)   '"Don't have a go at romantic
     fiction if you know nothing about sex"', Times
     Higher Education Supplement, No. 400 (27 June
     1980), p. 10
Anderson, M. (1981)   'Planning Policies and Develop-
     ment Control in the Sussex Downs A.O.N.B.',
     Town Planning Review, 52, 5-25
Berger, S.R. (1971) Dollar Harvest, D.C. Heath,
     Lexington, Massachusetts
Blacksell, M. and Gilg, A. (1977) 'Planning Control
     in an Area of Outstanding Natural Beauty',
     Social and Economic Administration, 11, 206-15
Bowler, I.R. (1979) Government and Agriculture: A
     Spatial Perspective, Longman, London
Burbach, R. and Flynn, P. (1980) Agribusiness in the
     Americas, Monthly Review Press, New York
Clark, G. (1979) 'Farm Amalgamations in Scotland',
     Scottish Geographical Magazine, 95, 93-107
Clark, G. (1981)   'Public Landownership in Scotland',
     Scottish Geographical Magazine, 97, 140-6
Cloke, P. (1979) Key Settlements in Rural Areas,
     Methuen, London
Crossman, R. (1975; 1976) The Diaries of a Cabinet
     Minister, Vols. 1 and 2, Hamish Hamilton and
     Jonathan Cape, London
Dear, M. and Clark, G.L. (1978) 'The State and
     Geographic Process: A Critical Review',
     Environment and Planning A, 10, 173-83
Department of the Environment (1974) Report of the
     National Park Policies Review Committee, Her
     Majesty's Stationery Office, London
Edwards, A. and Rogers, A. (1974) Agricultural
     Resources, Faber and Faber, London
Evans, A.F. (1969) 'The Impact of Taxation on
     Agriculture', Journal of Agricultural
     Economics, 20, 217-28
Found, W.C. (1971) A Theoretical Approach to Rural

Land-Use Patterns, Edward Arnold, New York

Gaskin, M. (1971) Freight Rates and Prices in the Islands, University of Aberdeen, and Highlands and Islands Development Board, Inverness

Geisler, C.C. (1980) 'The Quiet Revolution in Land Use Control Revisited', in F.H. Buttel and H. Newby (eds.),The Rural Sociology of the Advanced States: Critical Perspectives, Croom Helm, London, pp. 489-526

Gilg, A.W. (1978) Countryside Planning, David and Charles, Newton Abbot, Devon

Grafton, D. (1980)Planning for Remote Rural Areas, Discussion Paper 5, Department of Geography, University of Southampton

Hall, P. (1974) 'The Containment of Urban England', Geographical Journal, 140, 386-417

Healy, R.G. (1976) Land Use and the States, Resources for the Future, Washington, DC and Johns Hopkins University Press, Baltimore

Her Majesty's Stationery Office (1965) The Development of Agriculture, Cmnd. 2738, London

Huggett, F. (1975) The Land Question and European Society, Thames and Hudson, London

Jackson, R.H. (1981) Land Use in America, Edward Arnold, London

MacKay, D.I. and Mackay, G.A. (1975) The Political Economy of North Sea Oil, Martin Robertson, London

Mann, S.A. and Dickinson, J.A. (1980) 'State and Agriculture in Two Eras of American Capitalism' in F.H. Buttel and H. Newby (eds.), The Rural Sociology of the Advanced States: Critical Perspectives, Croom Helm, London,pp. 283-325

Ministry of Works and Planning (1942) Report of the Committee on Land Utilisation in Rural Areas, Cmd. 6378, Her Majesty's Stationery Office, London

Newby, H. (1979) Green and Pleasant Land?, Hutchinson, London

Northfield Committee (1979) Report of the Committee of Inquiry into the Acquisition and Occupation of Agricultural Land, Cmnd. 7599, Her Majesty's Stationery Office, London

Organisation for Economic Co-Operation and Development (1972) Structural Reform Measures in Agriculture, Paris

Organisation for Economic Co-Operation and Development (1974) Agricultural Policy in the United States, Paris

Pallot, J. (1979) 'Rural Settlement Planning in the U.S.S.R.', Soviet Studies, 31, 214-30

Poole, M.A. (1970) 'Decision Making and the
    Formulation of Regional Development Objectives -
    the Case of Rural Ireland' in N. Stephens and
    R.E. Glasscock (eds.), Irish Geographical
    Studies in Honour of E. Estyn Evans, Queen's
    University of Belfast, Belfast, pp. 312-24
Price, C. (1973) 'To the Future: with Indifference
    or Concern? - The Social Discount Rate and its
    Implications in Land Use', Journal of Agricul-
    tural Economics, 24, 393-8
Price, C. (1976) 'Blind Alleys and Open Prospects
    in Forest Economics', Forestry, 49, 99-107
Robson, N. (1971) An Evaluation of the Small Farm
    (Business Management) (Scotland) Scheme, 1965,
    North of Scotland College of Agriculture,
    Aberdeen
Rose, R. (1980) Do Parties Make a Difference?,
    Macmillan, London
Schuh, G.E. (1979) 'The Effects of Food Aid',
    Foreign Agriculture, 17, No. 13, 5-8
Smith, E.G., Jr. (1980) 'America's Richest Farms
    and Ranches', Annals of the Association of
    American Geographers, 70, 528-41
Tarrant, J.R. (1974) Agricultural Geography, David
    and Charles, Newton Abbot, Devon
Thomas, C. and Winyard, S. (1979) 'Rural Incomes'
    in J.M. Shaw (ed.), Rural Deprivation and
    Planning, Geo Abstracts, Norwich, pp. 21-46
US Chamber of Commerce (1974) The Changing Structure
    of U.S. Agriculture and its Contributions to
    the National Economy, Washington, DC
Wallerstein, M.B. (1981) Food for War - Food for
    Peace, Massachusetts Institute of Technology
    Press, Cambridge, Massachusetts
Whetham, E.H. (1960) The Economic Background to
    Agricultural Policy, Cambridge University Press,
    Cambridge
Wilson, G.K. (1975) The Policy Making Approach to
    British Agriculture in Open University,
    Agriculture, D203, Block III, Open University
    Press, Milton Keynes, Buckinghamshire, pp. 292-
    310
Wilson, G.K. (1978) 'Farmers' Organisations in
    Advanced Societies' in H. Newby (ed.), Inter-
    national Perspectives in Rural Sociology,
    Wiley, Chichester, pp. 31-53

Chapter 4
INSTITUTIONS AFFECTING ENVIRONMENTAL POLICY

Timothy O'Riordan

To define 'policy' is a difficult enough task but to characterise 'environmental policy' is much more treacherous. For there is no such thing as a coherent environmental perspective on issues of public importance nor is there likely to be an environmental logic to policy-making in the fore-seeable future. The US Government demonstrated this in the report Global 2000 (US Council on Environmental Quality, 1980) which reveals that the models of different federal agencies were not designed to be used together in a consistent and interactive manner. Indeed Global 2000 is being treated as a device to encourage greater collabora-tion among agencies with respect to an environmen-tally sustainable development strategy. Likewise the UK House of Lords Select Committee on the European Communities (1981a) concluded that the European Commission, the executive agency for the European Community, had developed no way of ascer-taining how far its sectoral policies and executive actions in areas such as regional economic develop-ment, agriculture, transport and energy took into account the various wide-ranging and long term effects on the environment. The Lords' Committee argued forcefully for the preparation of an inte-grated environmental strategy that should penetrate all relevant policy sectors, but one senses that for the foreseeable future this is wishful thinking.

One might well ask why a coherent environmental strategy has not been developed. The answer is simple: environmentalism cuts across established ways of tackling economic and social development. It challenges conventional approaches in economics, the law, politics, technological assessment and public attitudes to what is valued. Environmental-ism cannot be 'added on' to existing practices: it demands a fundamental restructuring of a whole philosophy and strategy. This is the message of the World Conservation Strategy (International Union for the Conservation of Nature, 1980) a powerful analysis of the global predicament that demands governmental response. Whether it will obtain that response before the chips are really down is quite another matter. A swing towards an integrated environmentally sustainable development strategy cannot come easily via incremental shifts in policy and practice but incrementalism is the name of the political game.

It is therefore too ambitious a task to review how institutions affect environmental policy in its proper, comprehensive sense. What will be attempted here is an analysis of how various kinds of institutions influence the outcome of certain aspects of environmentally related policies in four areas:

  (i)   environmental assessment (known as environmental impact assessment in the USA);
  (ii)   the safety evaluations of nuclear power plants (a narrow form of risk analysis);
  (iii) the control of air and water pollution (with some reference to noise and smell);
  (iv)   the protection of socially valued features of the countryside in the face of agriculturally induced change.

These four themes are chosen for illustrative purposes and in no way can they be said to encompass all aspects of environmental issues. But at least they encompass areas with which the author is familiar. An additional introductory caveat should be noted: in none of these areas will the substance of policies be discussed, for the purpose of this chapter is to analyse how institutions shape policy.

The Formulation of Environmental Policy

As already noted, the term 'policy' is an elusive one. It may refer to a set of guidelines or principles against which possible courses of action can be evaluated, or it may relate to a declared statement of intent to do something, backed up by the provision of an enabling budget. Institutions may prepare policy guides, but more often than not they embark on activities which by their very existence become policy. What eventually takes place therefore is rarely the consequence of a single analytical plan; it will have evolved from innumerable decisions taken by unknown individuals or by committees whose opinion may be deeply divided at various levels of influence and authority. As the liberal British independent 'think tank', The Outer Circle Policy Unit (1979, p. 31), put it:

> In Great Britain, 'policy' is not made at a kind of intellectual drawing board, where designers map out a complete and detailed scheme and then pass it on to others to put into practice. Instead policy is secreted in the interstices of individual decisions, and evolves over time as a consequence of the decisions that are taken and carried out. Too often, this happens without full consideration of all the implications...

One feature of the modern environmental movement is its challenge of existing sectoral policy-making procedures. The practice of preparing environmental impact statements (EISs) in the USA began largely as a device to force federal agencies to look at the broader societal implications of their actions and, further, to encourage them to co-ordinate their approach when major developmental decisions had to be taken. (For a comprehensive review of environmental impact assessment in the context of project assessment and the scrutiny of underlying policy, see O'Riordan and Sewell, 1981. For an excellent review of how US federal agencies responded to the dictates of the National Environmental Policy Act, see Ingram, 1976.) In the early 1970s this was particularly necessary in the USA because many administrative actions were ill-conceived, being promoted by powerful alliances of economic and developmental interests who had little genuine concern for the environmental repercussions. (For an excellent case study of the antics in the late 1960s of the US Corps of Engineers, today a much more environmentally sensitive organisation, see Drew, 1970.) In Europe, and especially in Britain, this 'environmental gap' to policy implementation was and still is rather less obvious because of established practices of extensive consultation with interested parties including various official and unofficial representatives of environmental bodies. However, nowadays these safeguards are no longer regarded as satisfactory. In short,

then, the environmental movement has insisted on and
increasingly won a recognition by developmental
agencies that they must formally and accountably
appraise the wider environmental and societal consequences of major courses of action.   And it is
this 'add on' element which becomes associated in
the minds of many people as environmental policy,
even though it is no such thing.

The Role of Pluralism

'Policy' is rarely consistent.   It is a reflection
of the competition of different interests, the outcome of the 'political struggle' as Wengert (1955)
put it.   Over the course of time these competing
forces will have greater or lesser power, hence
policy will be altered to suit the changing pattern
of political pressures.   This of course is the
pluralist perspective on policy-making as enunciated
by Dahl (1961) and as subsequently developed by many
theorists (for a good review, see Jenkins, 1978).
Pluralism depends upon two critical assumptions for
its credibility.   One is that interests are freely
able to form by coalescing around pivotal points of
common purpose.   The other is that they can gain
access to policy-makers before binding commitments
are made and moreover that they have a reasonable
chance of influencing events if their argument is
strong.   In practice these assumptions only
partially hold.   Some interests are politically
enduring and are always active on the political
scene.   In the context of this chapter two notable
British examples are the Confederation of British
Industry which has a lot to say about environmental
assessment and pollution control policies, and the
National Farmers,Union which is vociferously
defending its position over the agriculture-conserva-
tion conflict currently raging in the United
Kingdom (see Shoard, 1980).
    But the pluralist model breaks down where
issues do not attract a coherent interest group
capable of effective mobilisation, or where existing
interest groupsare divided over their views and
courses of action, or where the powers that be
deliberately conceal vital information until binding
political commitments have been made, at which time
the mobilisation of countervailing political pressure
is ineffective.   This kind of circumstance is most
commonly associated with the environmental component
of a policy, particularly in a country where secrecy

is the common currency of government, for the
environmental argument normally raises issues that
are problematic to established vested interests.
No industrialist is anxious to invest in pollution
control when there is no profit from the exercise,
except that of good public relations.   Likewise few
executive agencies relish the thought of undertaking
exhaustive environmental assessments because these
are time consuming and expensive to prepare and they
open up avenues of information which mischievous
objectors may happily exploit.

The Role of Accommodation and Containment

Thus many features of environmentally related policy
are more suitable for the 'non-decision-making'
treatment first discussed by Bachrach and Baratz
(1962) and developed by Jenkins (1978).   The
practice of non-decision-making is a little tricky
to describe for it refers to the deliberate and
successful practice of keeping a potentially contentious issue away from public attention and
political discussion.   Crenson (1971) found this to
be the case with respect to pollution control in
Gary, Indiana. Neither the US Steel Corporation nor
the local pollution control officers wanted their
carefully orchestrated negotiations to become too
public so they protected each other's confidence and
reached an accommodation over acceptable emission
levels which may have been to their mutual satisfaction but which could by no means be regarded as a
reflection of the public interest as defined by
Wengert.   Accommodation and containment therefore
become important features of the modern environmental
game: accommodation amongst mutually understanding
interests and containment from the gaze of the
potentially 'meddling' public.   This is certainly
the case in Britain and France with respect to such
items as pollution control and risk/safety assessments where official secrecy and a faith in the
sound judgement of officials and their expert
advisers provide a powerful protective argument.
In the more open governments of the US and Sweden
and, most recently,Canada containment may be less
feasible (though rarely impossible) though accommodation will still be an important objective.

Timothy O'Riordan

The Policymaking Culture

The policy-making culture comprises the set of
traditions, conventions and expected ways of doing
things which drive the machinery of government.   A
policy-making culture is essentially a national
concept for it relates to the legacies of national
administrative procedures and institutional arrangements and to what people operating in nation states
regard as customary behaviour.   This culture varies
greatly from nation to nation but the gulf is particularly noticeable when comparing British and
American experience.   Much of this has to do with
constitutional history.

The US Experience

The American founding fathers devised a form of
government that was mutually antagonistic and in
which no single centre of power could predominate
over any other.   Thus the three principal arms of
US government - the Congress, the Administration
and the Judiciary - are separate but equal and each
has a vital executive role to play in the formulation
of policy (see Figure 4.1).   Furthermore the
Americans founded a system of government based on a
sense of presumed mistrust of officialdom, so
various devices are available for making public
officials accountable for their actions.   These
include requirements to document all the reasons
for adopting a particular course of action, to
appear before committees of investigation when
requested to do so, and to be answerable before the
courts if necessary.   As a result, US policy-making
procedures tend to be paper heavy and legally
complex involving extensive use of the photocopying
machine and legal counsel.   Since lawyers often
act as intermediaries between principal negotiating
parties it is not surprising that the US environmental lawyer is nowadays a highly trained and highly
paid individual who may prove to be most influential
in determining the outcomes of environmentally
related policy issues.
     In the USA too lobbying is a recognised
political art which has been adopted by the major
environmental organisations such as the Sierra Club,
the Audubon Society, the Natural Resources Defense
Council and the Environmental Defense Fund in imita-
tion of their 'enemy', the industrial and resource

108

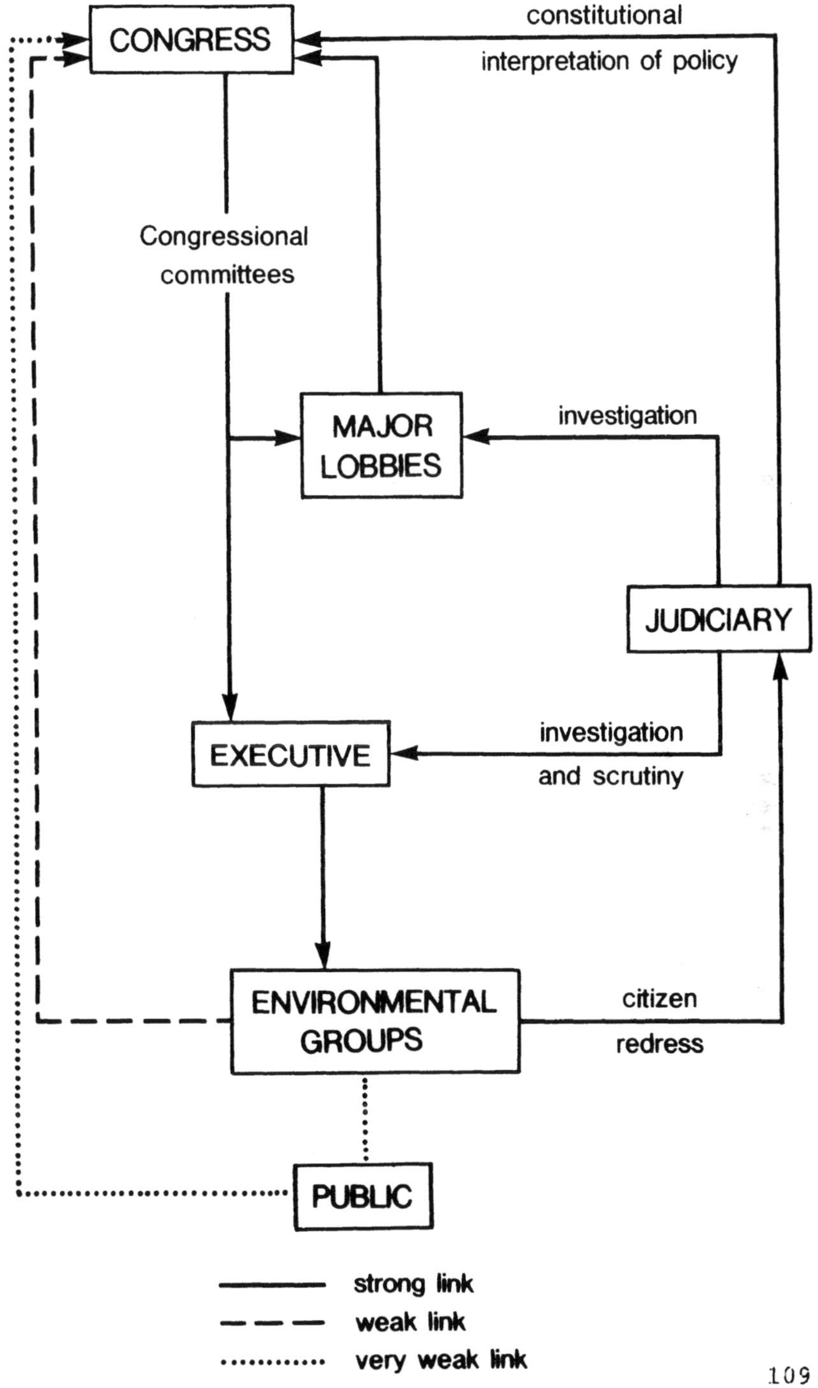

Figure 4.1: Constitutional Relationships within the US Government

exploitation groups. Lobbying means more than the peddling of influence, it means currying favour by various legal and illegal means, preparing briefing papers for politicians and administrators, attending important public meetings and conferences and above all knowing in advance when, how and upon whom to act. Lobbying is a game in which the stakes are high. A timber company may lose many millions of dollars of potential revenue if an area in which it has a commercial interest is declared a wilderness and thus exempt from logging. Likewise an electricity utility corporation may have to spend many millions to meet new nuclear safety standards, often by reconstructing existing equipment, should tough regulations be promulgated. In many respects it is the cut and thrust of countervailing lobbying which shapes an environmental policy, but again one should beware of adopting a purely pluralist perspective. So much has to do with the particular characteristics of the issues in contention, the personalities and world views of the key actors involved, the degree of advance intelligence available to the participating parties, the extent to which constituency pressures can be brought to bear on individual politicans and the translation of these pressures to governmental action.

The UK Experience

In the UK the conventions are very different. There is no written constitution so much depends upon the interpretation of tradition and precedent. Parliament is the supreme policy-making body, to which, in theory at least, the ministerial departments and the courts are subservient (see Figure 4.2). There are few countervailing constitutional checks to the law of Parliament, which is answerable only to the electorate. A number of observers have commented that Britain now suffers from an 'elected dictatorship' in which a democratically selected Parliament rules supreme and virtually unrestrained for the period between general elections. Many distinguished British lawyers argue passionately for greater controls on the authority of Parliament and a legally enshrined Bill of Rights (see Scarman, 1974).

Figure 4.2: The Institutions of Policy-making in
Britain and their Relationships

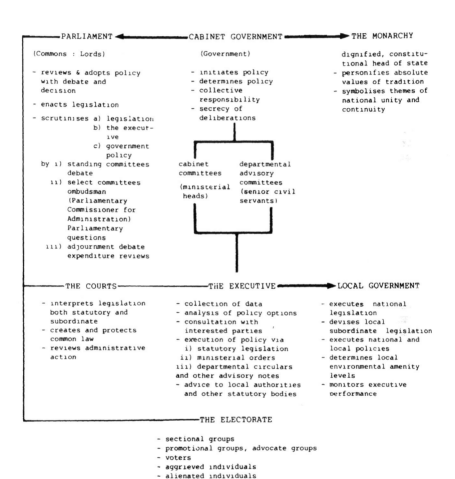

PARLIAMENT ◄──────── CABINET GOVERNMENT ────────► THE MONARCHY

(Commons : Lords)   (Government)

- reviews & adopts policy
  with debate and
  decision

- enacts legislation

- scrutinises a) legislation
  b) the execut-
  ive
  c) government
  policy
  by i) standing committees
  debate
  ii) select committees
  ombudsman
  (Parliamentary
  Commissioner for
  Administration)
  Parliamentary
  questions
  iii) adjournment debate
  expenditure reviews

- initiates policy
- determines policy
- collective
  responsibility
- secrecy of
  deliberations

dignified, constitu-
tional head of state
- personifies absolute
  values of tradition
- symbolises themes of
  national unity and
  continuity

cabinet | departmental
committees | advisory
(ministerial | committees
heads) | (senior civil
| servants)

THE COURTS ──────── THE EXECUTIVE ────────► LOCAL GOVERNMENT

- interprets legislation
  both statutory and
  subordinate
- creates and protects
  common law
- reviews administrative
  action

- collection of data
- analysis of policy options
- consultation with
  interested parties
- execution of policy via
  i) statutory legislation
  ii) ministerial orders
  iii) departmental circulars
  and other advisory notes
- advice to local authorities
  and other statutory bodies

- executes national
  legislation
- devises local
  subordinate legislation
- executes national and
  local policies
- determines local
  environmental amenity
  levels
- monitors executive
  performance

THE ELECTORATE

- sectional groups
- promotional groups, advocate groups
- voters
- aggrieved individuals
- alienated individuals

Timothy O'Riordan

The Role of the European Community

So far the discussion of Britain has centred on
national government. This is only one of the
relevant levels of governments, the others being
the supranational European Community and the
plethora of local governments. Britain has always
had an uneasy relationship with the European
Community, a coalition of ten member states that
determines a surprising number of policies and
actions across a wide range of activities. Indeed
there is still active political debate within
Britain as to its future role vis-à-vis the
Community. The Community influences national
environmental policies in two ways, first by
issuing directives which are (in theory at least)
legally binding and, second, by promoting policies
which involve substantial transfers of money and
which lead to actions which may have serious environmental consequences.

The most notable group of directives are those
dealing with environmental assessment and pollution
control. The European philosophy is to have
common regulations and harmonised practices so that
no member state can avoid its environmental responsibilities and thereby gain competitive
advantage. So the Community has issued directives
on water quality, on air quality (especially with
respect to sulphur dioxide) and on certain toxic
substances, all of which are quite precise in their
requirements and all of which are based on the twin
principles of strict emission standards and ambient
quality (i.e. general air and water quality) guidelines (see O'Riordan, 1979a). To date British
governments have refused to accept the principle of
fixed emission standards irrespective of the
variable capacity of the local air and water flows
to absorb pollutants and have forced the Community
to accept the establishment of general air and
water quality guidelines as targets to which action
must be directed. For its part the Community has
forced the UK to adopt a more stringent approach to
pollution control. With respect to environmental
assessment, the British have steadfastly refused to
accept the principle of a mandatory requirement on
all projects and programmes significantly affecting
the environment and so far have managed to fend off
the relevant directive which is now in its twenty-
first draft.

European Community policies in areas such as

agriculture, transport and regional development are far more threatening to environmental quality yet less subject to criticism from successive British governments. For example the European Community has recently succeeded in persuading the British to accept heavier lorries (from a maximum of 32 tons up to 44 tons) (Armitage, 1980) despite tremendous protest from the amenity lobby that these juggernauts would cause unacceptable noise and damage the pleasantness and fabric of some of the country's ancient towns.

Equally controversial are the environmental effects of the Community's Common Agricultural Policy (CAP). This provides for a major injection of money to prop up prices for a great array of farm produce plus subsidies to aid farm improvement schemes. The result is a massive intensification and capitalisation of agriculture with all kinds of associated effects on traditional landscapes in upland and wetland areas, significant loss of natural habitats due to the use of chemical fertilisers and pesticides and the emergence of a new and revitalised farming lobby with very high expectations of its continued economic viability. In a similar vein regional development funds provided by the Community inflame local optimism that money will be available for a variety of schemes ranging from motorways to power stations, many of which are opposed by local environmental groups.

One should not underestimate the effect of Community decisions on national environmental policies. While the conflict over directives is easier to discern the environmental implications of Community strategic policies are more difficult to identify and almost impossible to counteract. Sectoral policies, in whose creation a particular member state may play only a small part, can become so pervasive and can establish so many powerful expectations amongst influential lobbies that they can become politically fossilised. Yet they may establish conditions which can lead to a serious and progressive deterioration of environmental quality over which there may be virtually no political accountability. This is particularly the case at present with respect to CAP where the British government is struggling to identify the proper balance between agricultural improvement and countryside protection and conservation only to discover that the major culprit is neither its farmers nor its own uneven pattern of subsidies and incentives.

The principal blame rests with the CAP itself which
subsidises farmers to destroy rural environments
by a vast array of price support schemes. Yet a
single national government is unlikely to alter
this arrangement in any substantial manner.

The Role of Local Government

In Britain, 'local government' refers to the
executive agencies and decision-making bodies
associated with county councils and district
councils created under the Local Government Act of
1972. While in theory these bodies are responsible
for a variety of key policy areas - housing, transportation, planning and pollution control, education
and social services, in practice their budgets are
in large part determined by central government dis-
bursements (known as the Rate Support Grant). This
means that unless these bodies are prepared to
raise local rates (taxes), always a politically unpopular move, their policies are shaped by budgetary
stringency. The present Conservative government
in Britain is particularly anxious to implement a
major element of its economic strategy, the control
of public spending, through this means. So in
terms of policy formulation local governments are
more restricted than they appear constitutionally.
We shall see how fiscal restraint imposed by
central government also affects the implementation
of policy at a local level, particularly in the
area of pollution control and environmental assessment where the major responsibility lies with
officers working in the county and district
councils.

   One feature of environmental assessment that
needs more discussion here is the imbalance of
costs associated with a major public sector development proposal such as an airport, a power station,
a chemical complex or a mining scheme. In every
case the local authority (normally the county but
also the relevant district council) is faced with
large costs associated with financing the infrastructure (roads, schools, libraries, housing) linked
to major schemes. While some of these costs are
taken into account through modifications to the
Rate Support Grant, not all are. It is now
evident that the local ratepayer has to pay not just
for the environmental disamenities resulting from
these schemes, disamenities which may be considerable
but which are difficult to calculate and to

compensate for, but also for the additional infrastructure services many of which he is not particularly interested in enlarging.   It is thus little
wonder that both the County Councils Association and
the Association of District Councils have made
strong complaints to central government to redress
this imbalance.   To date they have not been given a
sympathetic hearing.   The result is that most
counties and districts oppose major development
schemes, especially those promoted by the public
sector.   This in turn can trigger off a demand for
a full scale public inquiry and with it an opportunity to comment on the underlying assumptions of
government policy.

The Concept of 'Government'

This distinction between the US and UK governmental
systems is an important one to make because it
reveals that the notion of 'government' differs from
one country to another.   In the US 'government' is
a polycentric notion where power is spread over a
number of nodes - congressional committees, the
offices of senior administration officials, the
lobbies of key interest groups, the White House and
the courts.   Government is a concept that breeds a
sense of mistrust but is believed to be penetrable
and manipulable if one is determined enough.   In
the UK 'government' is formally equated with the
Prime Minister, his/her cabinet and the phalanx of
junior ministers, political aides and senior civil
servants who advise and execute policy.   This nexus
of power is remarkably monocentric, is mostly
impenetrable due to the rules of confidentiality and
the laws of official secrecy, and is regarded with
an air of resigned suspicion by outside parties
seeking entry.   Parliament is by no means
'government' in Britain: it is a theatre in which
the mood of the nation is translated into political
rhetoric.   A party in power with a comfortable
voting majority can always control Parliament,
almost irrespective of public discontent with a
particular course of action.   The only danger to
the government is the threat of a backbench revolt
by members of the ruling party: such a revolt is
extremely rare on the really important matters of
state.

Thus in Britain the notion of government is a
vague and often misdirected one which is undergoing
reassessment.   In the past the concept of government

has been associated with a paternalistic feeling, a belief that 'nanny knows best' and that those chosen to rule and to advise could be trusted to recognise and implement the public interest. Hence the practice of secrecy whereby nobody is accountable except the responsible politician at the top and he decides what supporting documentation and underlying assumptions to reveal. In addition British government depends a lot on the flexibility of discretion, the freedom to use judgement unencumbered by strict laws and rules. Discretion pervades both political and administrative practice and is very much criticised by environmental pressure groups who regard it as a means for avoiding formal accountability.

In the field of pollution control, for example, ministers have wide powers to set and to alter standards of 'acceptable' nuisance and administrators (regulatory officials at various levels of government including local government) have equal flexibility to determine what levels of emission are tolerable. The convention here is the 'best practicable means' principle, through which pollution control officials bargain with polluters. They do so usually from a position of relative weakness since the discharger has all the relevant information on control technologies and how they can be applied to his circumstances, on the costs and on his own economic viability. Bargaining takes place over criteria such as the levels of nuisance already in existence in the locality in question (and the interpretation of nuisance will vary according to the income of the residents, their associated expectations and property values), the technological options available to the discharger for reducing the noxious elements of his emission, the economic circumstances of the polluting activity (i.e. how much the proposed pollution control will cost in relation to profitability) and any other relevant and material considerations that the bargaining groups may choose to take into account.

Here clearly is where the practices of containment and accommodation, alluded to earlier, come into play. Combine this flexibility with the view held by regulators (including planners involved in environmental assessments) that they are professionals who are able to advise their clients as much as to control them, and with an unwillingness to enforce compliance through prosecution and/or a demand for a public inquiry, both of which require the public exposure of an argument and can fracture the client-regulator relationship, and one can readily see why environmental regulation in Britain is regarded with so much suspicion by informed members of the public and why accommodation and regulation become key elements in the political game.

Hence, in Britain, the watchword is accountability. The intent of most environmental lobbies (together with other watchdog groups) is to break down the barricades of confidential advice and discretionary flexibility so as to ensure that all decisions are taken out in the open with full supporting documentation suitable for cross examination. This is why, in this chapter, attention is given to a number of key institutions established to review policy and to scrutinise decisions - the legislative committee, the public inquiry, the specialist advisory bodies (known as quangos) and the courts. It is this mix of institutions, all of which are subsidiary to 'government' in its constitutional sense, which environmental activists perceive as providing them with means to ensure greater accountability, or, to be more accurate, greater responsiveness to their demands.

In the USA, this obsession with accountability is less noticeable, because in constitutional theory at least accountability is an established fact. There the interest lies in how lobbying influences accountability and how administrative budgets and thus the calibre and number of administrative personnel are allocated in respect of the tasks expected of the administrative agencies. There is little point in setting up precise regulations and demanding administrative procedures if agencies are ill equipped to execute them. Currently this seems to be a bit of a problem with respect to environmental assessment since a number of federal agencies are inadequately financed to carry out all of their duties under the National Environmental Policy Act of 1969 (NEPA) while the Council on Environmental Quality, the overseeing agency in this critical area, has recently suffered a major reduction in budget and manpower. One senses that a new round of environmental battles is just about to begin in the USA with an impressive marshalling of forces on each side.

Institutions for Reviewing Environmental Policy

For the purposes of this chapter we will assume that there are four principal institutions designed in

some manner to review environmental policy. These
are the legislature, the public hearing, the advisory
committee and the courts. As noted earlier each
plays a different constitutional role depending on
the national context and the matter to hand.

The Legislature

In the USA, Congress reviews policy primarily through
the medium of its specialised committees. These
are very powerful bodies, chaired by experienced and
senior politicians and supported by large research
staffs and legal advisers. Generally speaking
committees select issues for analysis according to
events, to pressure from the lobbies and the
Administration, and to a sense of the public mood.
The committee hearing is an elaborate affair
accompanied by extensive documentation and much
cross-examination. Hearings may be held on problematic issues, such as a particular disaster or set of
regulations, or they may focus on proposed legislation. The hearing forms the basis of a committee
report, itself the result of much lobbying and
amendment, and this report, together with its
counterpart from the other Congressional House form
the basis for further negotiation and compromise.
The outcome of this, assuming that some common
agreement can be reached, is congressional policy.
This document then forms the basis of further discussions with the White House, the result of which
should be national policy. However, interested
parties can always test the constitutionality of
this policy in the courts.

The congressional committee is a powerful policy-
creating as well as policy-reviewing institution.
Once it has come to terms with the evidence and
agreed a common negotiating position with all its
politically influential partners, then this is
policy. A good example of the policy-creating role
of the congressional committee is that of the Senate
Interior and Insular Affairs Committee under
Senator Jackson which masterminded the National
Environmental Policy Act. This particular piece of
legislation not only mandated the environmental
impact statement in all federal agencies but it
established the vital principle that economic
development should take place in 'productive harmony'
with the protection and enhancement of environmental
quality. No other single piece of legislation anywhere in the world established such a precedent for

environmental policy-making yet curiously neither the
Jackson committee nor its counterpart in the House
of Representatives fully understood the significance
of their statutory offspring (see Anderson, 1973,
and Dreyfus and Ingram, 1976). Congressional
committees have remarkable freedom of action: they
can initiate a study, they can promote a bill of an
individual congressman, they can respond to a powerful lobbying effort. They are not dictated to by
the Administration and their reports, once internally agreed with counterpart committees, often
become binding on both Houses.

By contrast, the British parliamentary equivalent, the select committee, is a much tamer affair
both in terms of creating policy and as a policy-
reviewing institution. The reason is that,
constitutionally speaking, Parliament has a limited
role in reviewing policy and, until recently,
Parliament as a whole did not take the reports of
its select committees very seriously. The British
Parliament is a remarkably conservative institution
and many of its long-serving members genuinely do
not believe in specialised committees and the
generation of expertise amongst their fellows.

As a result, Parliament is largely regarded as
a debating chamber and not a policy-reviewing institution. Indeed there have been few informed and
lengthy debates on any aspect of environmental
policy over the ten years since the environmental
movement gained respectability. The reason for all
this is once again constitutional and traditional.
Policy is formulated within departments of state,
not by individual legislators,it is promoted by
ministers either through Cabinet or directly through
their departments and there is little opportunity
for policy to emerge from any other source. How
policy is created within a department of state is
still somewhat of a mystery for the process is
deliberately made cryptic. Figure 4.3 offers one
suggested pathway where the role of policy-
co-ordinating groups in the middle levels of the
civil service hierarchy is of particular significance. The pathway is based on extensive consultation with key interested parties, a sense of the
political mind of the minister in charge and an
awareness of the general public mood on the issue to
hand if such a mood can be identified. In such
cases Parliament is mostly a policy-reactive rather
than a policy-reviewing institution.

Since 1979, however, the British Parliamentary
select committee system has experienced an important

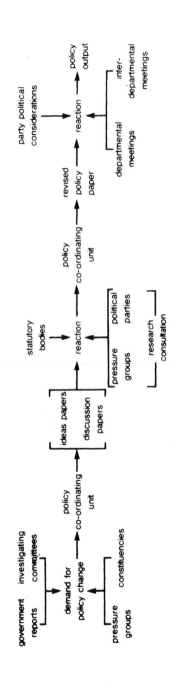

Figure 4.3: Environmental Policy-making in UK Central Government Departments

renaissance.   Now there are 12 special committees
(see Table 4.1) each responsible for looking at the
affairs of a single or related group of ministerial
departments.   Their job is to take any aspect of
departmental expenditure and/or policy and assess
its relevance and cost effectiveness carefully.
The result has been a spectacular shift in power
toward the back benches, a much greater parliamentary
respect for its select committees, and an improvement in the level of expertise amongst members.

    The great value of the new breed of departmental
select committees lies in their freedom to choose
the subject of their investigation, their power to
cross examine witnesses, including ministers, and
their assembly of memoranda and other evidence
which collectively provide an invaluable source of
information.   The modern select committee gives an
issue full exposure thereby providing plenty of
evidence for interested parties to argue their
positions.   Their reports are now more likely to be
debated in the House during which the ruling party is
expected to make its position vis-à-vis the issue in
question quite clear.   So the select committee plays
a significant catalytic role in bringing contentious
issues onto the public agenda.   A further feature
of the modern select committee is the better research
support available together with the employment of a
number of specialist advisers.   Their job is to
assemble the evidence, to guide the committee chairman as to who should be examined and what questions
should be asked, and to prepare the drafts of the
final report in the process of which they are
expected to read the committee members' minds.

    Two recent examples illustrate the role of the
Parliamentary select committee.   In December 1979
the Secretary of State for Energy announced that the
Government accepted the projections by the electricity supply industries that one new nuclear
power station should be built every year for a
decade to meet anticipated electricity needs.   The
Energy Secretary also indicated that the Government
wished that at least one of these stations be an
American designed pressurised water reactor (PWR),
subject to satisfactory safety clearance by the
Nuclear Installations Inspectorate, the official, but
only quasi-governmental, nuclear regulatory agency.
The Select Committee on Energy (1981) then proceeded
to take evidence on both the likely cost of a PWR
and its safety.   Its choice of advisers reflected
these twin themes of investigation.   In the event
the Committee was most unhappy about the proposed

Timothy O'Riordan

Table 4.1:   The New Policy-related Parliamentary
Select Committees

Name of Committee	Government Departments Covered
Agriculture	Ministry of Agriculture, Fisheries and Food
Defence	Ministry of Defence
Education, Science and the Arts	Department of Education and Science
Energy	Department of Energy
Environment	Department of Environment
Foreign Affairs	Foreign and Commonwealth Office, Overseas Development Administration
Home Affairs	Home Office, Lord Chancellor's Department, Law Officers' Department
Industry and Employment	Department of Industry, Department of Employment
Social Services	Department of Health and Social Security
Trade and Consumer Affairs	Department of Trade, Department of Prices and Consumer Protection
Transport	Department of Transport
Treasury	Treasury, Civil Service Department, Parliamentary Commissioner for Administration

cost estimates presented by the electricity utility
(the Central Electricity Generating Board) and
requested that it recalculate its estimates.   In
its final report it argued that there were too many
uncertainties surrounding the cost forecasts to
allow it to conclude that the PWR really was the
least expensive nuclear option.   Since then newspaper reports indicate that the PWR costs may be at
least 50 per cent higher than suggested by the
Board during the committee hearings.   If true, this
would undermine the Board's case for the PWR which
is largely based on its cheapness relative to other
reactor designs.   The Committee also recommended

that, before any new nuclear plant was proposed, a
full cost audit be prepared coupled with a detailed
analysis as to its need vis-à-vis electricity demand
projection and alternative supply options.   If this
recommendation is accepted by the Government, it
would open up a wide avenue for critical opposition
by the anti-nuclear lobby.   The Committee also took
evidence on the controversial matter of reactor
safety from a number of informed witnesses.   They
concluded that the PWR could be made acceptably safe
but only under the most exhaustive and stringent
supervision and that the Nuclear Installation
Inspectorate should be better staffed and equipped
to handle the safety analyses.   The Committee's
caution over the safety issue was duly noted by the
anti-nuclear lobby which is already preparing its
case to oppose any PWR proposal (Friends of the
Earth, 1981).

     The second example concerns a Lords' select
committee report on environmental assessment (EA)
(House of Lords, Select Committee on the European
Communities, 1981b).   This committee took evidence
on the European Community draft directive on EA
during which it heard a wide range of views for and
against the mandatory and systematic introduction
of EA for major schemes.   The main point to note
here is that the evidence from senior officials in
the Department of the Environment, which was all but
the same as the official governmental position (in
itself an interesting revelation), did not favour
the introduction of EA except on a voluntary basis
to be prepared by the promoting agency according to
particular circumstances.   Here is the classic re-
statement of voluntary flexibility based on
ad hocery.   Needless to say the Confederation of
British Industry supported this position as did
senior planners who feel that existing safeguards
are satisfactory.   But the younger planners and
the environmental activists did not endorse this
rather smug approach and persuaded the Committee
that the directive with few minor modifications
should provide the basis for a new policy.   Here
was a senior Lords committee opposing established
government policy having reached its conclusion on
the basis of wide ranging evidence.   An outcome
like this heartens the environmental organisations
who are encouraged to lobby more intensively.   Such
is the perceived influence of the Parliamentary
select committee.   In the event, however, the
government reiterated its opposition to the Community
directive in the Lords' debate on the committee

report and a subsequent Commons' committee is
apparently likely to follow a similar line.

These two case studies illustrate the merits and
drawbacks of the select committee as a policy-reviewing institution.   The advantages lie not so much in
substance as in procedure.   The conclusions of a
committee will be handled by the political mill
according to prejudice but the evidence, articulated
analysis and succour given to various groups
associated with a committee's inquiry all help to
strengthen the hand of the environmental lobby no
matter what the final outcome.   The reformed select
committee structure is now part of the political
fabric in Britain: its role as an environmental
policy-reviewing institution is likely to become
more influential and demanding over the next decade.

The Public Inquiry

So far we have discussed a policy review institution
that is run by legislators, who, constitutionally
speaking, should be the people for the job.   But
because of disenchantment, especially in Britain,
over the inadequacy of this role, many leading
environmental groups have sought an extra-
parliamentary forum for reviewing policy where they
have direct inquisitorial access.   This forum is
the public inquiry, a statutory device run on quasi-
judicial lines through which opponents and proponents
of controversial development proposals may argue
their case before an impartial chairman or tribunal.
In Britain the public inquiry is an important aspect
of what is known as natural justice, the philosophy
that all sides have a right to take their case fairly
and fully before an independent judge before a final
political decision is made.   In the USA, the equivalent is the public hearing which is a less formal
affair though broadly speaking it is based on
similar principles.

The major difficulty with the public inquiry as
an institution for reviewing environmental policy
lies in its variability of purpose and fickleness of
occurrence.   In the case of a controversial development proposal such as a new power station or mining
development, the application for planning
permission is normally lodged with the local government planning authority.   (For the power stations,
the Secretary of State for Energy may become involved
while major drainage schemes are dealt with by the
Minister for Agriculture, Fisheries and Food.   The

124

public inquiry and its statutory local planning inquiry are not quite the same thing, though for major developments the one merges with the other.) Should this authority be minded to approve the proposal then no public inquiry need be held unless the relevant minister feels that the application should be 'called in' before an inquiry in the public interest. So the inquiry is a bit of a hit and miss affair and if a promoting body does its homework well and prepares voluntarily a full environmental assessment of its scheme then it may avoid the time consuming and costly inquiry. It is hardly surprising to record that many of the major developmental organisations now have environmental assessment units attached to their managerial teams with a view to 'smoothing the path' of a development application.

Apart from the doubt whether an inquiry will actually be convened in a given instance, the other major problem with the public inquiry is that it is not designed to debate the policy that underlies a proposal, only the merits or otherwise of the particular application before it and particularly its effects on the locality. So, in theory, an inquiry inspector can rule out evidence on whether the scheme is needed at all, whether the policy relating to the proposal is sensible or not, and even whether alternative schemes or locations should be considered. In practice, inspectors have allowed some evidence on these themes to be heard, but normally they do not comment on these matters when writing up their report. The government has also made it clear that it does not expect the inquiry to be a policy-reviewing institution and presumably it has instructed its civil servants not to permit themselves to be cross-examined on such matters.

The major UK environmental lobbies are becoming very frustrated about all this and very suspicious of the whole inquiry procedure. They tend to see it as a window dressing exercise protecting a governmental commitment that may already have been made. They are jaundiced about the genuine impartiality of the inspectors who are appointed by the Lord Chancellor's Office but who are seen as part of government, and above all they are annoyed that no public money is available to help finance their participation when they are faced by development proposals paid for in the case of public corporations by taxpayers' money. This is a particularly annoying matter as the ratio of

125

expenditure between the promoting body and protesting groups can be as high as 5 to 1.   Again the present Government position is uncompromising: if pressure groups wish to oppose and feel strongly enough to do so, then they should raise their own money.   But in a case such as the proposed PWR plant where a major safety case has to be reviewed, the cost of a full scale independent analysis could be £50,000 together with another £25,000 to present the case using legal counsel at a public inquiry.   Meanwhile the nuclear industry and the electricity generating board which are mostly financed by taxpayers and electricity consumers will spend millions of pounds in preparing its full safety case.

So the public inquiry is under fire in Britain and it looks as if it will be taking considerable strain over the next few years as major battles shape up over nuclear energy, the proposed third London airport at Stansted and land based oil and gas deposits along the south coast.   Frankly, it is ill-designed for the task it should now be performing.   The great British traditions of ad hocery and infinite flexibility may be sorely tested unless some sense prevails and the inquiry is modified to show that it is an unprejudiced exercise where fairness of procedure is not compromised by a genuine lack of resources and hence suitable expertise.

The Quango

A simple and relatively inexpensive method for the government to appraise the merits of possible options relating to environmental issues is to establish an advisory body of some kind.   These range from full-scale civil service departments, such as the Countryside Commission and the Health and Safety Executive to highly professional bodies of scientists, as in the case of the Nature Conservancy Council, to committees of various sizes composed of experts often working in a voluntary capacity.   The all embracing term for these institutions is the quango - the quasi-autonomous non-governmental organisation.   Quangos abound on the British political and administrative scene because they are an outcome of the traditions of voluntary public service and selective consultation amongst informed interests.

Though technically independent of government quangos are not independently powerful.   They fall midway between the select committee and the public inquiry but are less accountable than either and often their mode of working, their recommendations and their existence are unknown to the public. For better or worse quangos are creatures of the system and as such they are exploited by the system sometimes to justify an unpopular decision already made behind the scenes, sometimes to let the government off the hook when it is faced with a particularly difficult problem. Members of quangos are at times placed in a dilemma. They may not have sufficient information or be expert enough to pronounce definitely on a matter where they feel that action should be taken. Or they may choose to adopt a stance only to find that it is unacceptable to either the government in power or particular pressure groups who may have enough clout to make political trouble. In neither case however are they individually or collectively accountable which, naturally, renders them the focus of attention of any major lobby, environmental or otherwise, when their recommendations are controversial.

One of the quangos most important to British environmental policy is the Royal Commission on Environmental Pollution. This body was established in 1970 as a response to the Torrey Canyon disaster of 1967 when it was recognised that there should be greater co-ordination amongst scientific and administrative departments with respect to pollution control. The Commission has always been led by an extremely able chairman and has been staffed by very competent researchers. It has taken evidence on a variety of topics and has shown itself to be quite fearless about probing into sensitive areas of environmental policy. For example in its Sixth Report on nuclear power (1976b) it made it clear that it was unhappy about the rapidity with which the 'plutonium economy' based on a fast-breeder reactor was being promoted. Its caution encouraged anti-nuclear groups to redouble their efforts to oppose the application for a thermal oxide reprocessing plant at Windscale (Wynne, 1980) and in part helped to slow down the rate of nuclear expansion in Britain.

The Commission's Fifth Report (1976a) on air pollution was probably its best in terms of environmental policy analysis. In that investigation the Commission supported the flexibility of the 'best practicable means' approach but argued that it should be applied to a comprehensive pollution control policy coupling air and water emissions and renamed the 'best practicable environmental option'. The Commission further believed that the local and

Timothy O'Riordan

central government pollution control authorities
should be more closely linked with a unified pollution inspectorate dealing with the more problematical
activities.   Finally the Commission endorsed the
principle behind the European Community's view of
ambient air quality objectives and recommended that
'target bands' of desirable air quality should form
the basis of a future emission control policy.
    These were all powerful and radical suggestions
and showed the Commission to be a significant force
in the politics of pollution control.   However, the
recommendations were unfortunately too radical both
for the regulatory agencies and for the Confederation of British Industry who favoured the status quo.
Even though it is now five years since the
Commission reported, the Government has still not
yet officially responded to its recommendations.
This indicates one of the dilemmas of the quango -
it may be controversial, it may stimulate debate and
encourage interests to mobilise their arguments and
it may certainly generate a most useful body of
information - but it has no power to alter policy.
Much will depend on the political characteristics of
the issues under review, the amount of information
already available, the personalities and status of
the quango members, especially the chairman, and
the degree to which the media choose to take up its
cause, particularly if a parliamentary debate on the
matter is in the offing.   It would perhaps be too
harsh a judgement to say that quangos are exploited
for wider political purposes but certainly there is
an element of manipulation about their position in
the policy-generation process, their lack of formal
accountability and the manner in which their
recommendations can be used.

The Courts

In the USA the Judiciary is an arm of government
equal with the Administration and Congress.   The
courts not only have a powerful role to play in
reviewing any executive action to ensure its
legality and due process, but they may also be used
to test the constitutionality of legislation.   From
the environmental point of view, however, the
principal role of the courts is to ensure that
executive agencies do what they are supposed to do,
that they follow the intent of the procedures that
are so specifically laid down for them and that they
show that they are balancing interests when required

to do so.   The law therefore provides an avenue for
testing both administrative competence and, by
implication, the quality of the arguments of the
environmental lobby.

     The trigger to the recent spate of environmental
litigation in the USA was the passage of the Administrative Procedures Act of 1967 which provides that
'A person suffering legal wrong because of agency
action, or (who is) adversely affected or aggrieved
by agency action within the meaning of a relevant
statute, is entitled to judicial review thereof.'
This provision enabled public interest groups to
appear before a court where they felt the public
interest might be harmed by a forthcoming agency
action or where they believed that an agency was not
following administrative procedures as laid down by
Congress or agreed through other means.   The Act
was passed at a time when US tax legislation encouraged the formation of public interest law
groups by granting them charitable status and thereby
enabling them to solicit tax-deductible donations
from benefactors.   By 1970 many of the major US
environmental bodies had established legal arms and
a number of new bodies were formed specifically with
a view to promoting the cause of environmental
litigation (see Table 4.2).

     During the early 1970s many US federal agencies
passed through a fairly painful period of learning
to accommodate to environmental groups.   The key to
the success of these groups was the passage of the
National Environmental Policy Act in 1969 which
required federal agencies to prepare environmental
impact statements on all actions significantly
affecting the human environment.   The environmental
groups used the courts to ensure not only that the
agencies heeded this statute procedurally but also
that they shaped policies and specific decisions to
show that they had taken environmental safeguards
into account.   This transition was successful
largely through the medium of litigation.   During
the first ten years of NEPA some 12,500 EISs have
been prepared of which 1,186 involved court action
and some 220 led to court imposed injunctions on
some aspect of the proposal.   There are some
sceptics who believe that NEPA has not substantially
changed anything and that the EIS has become a
matter of form rather than substance (see Fairfax,
1978) but most observers accept that agency planning
and decision procedures have been improved as a
result of NEPA and that the recent requirements for
agencies to show how they have taken an EIS into

Timothy O'Riordan

Table 4.2: Public Interest Environmental Groups in
the United States

Name	Date of founding	Membership
Established Groups		
Sierra Club	1892	178,000
National Audubon Society	1905	373,000
National Parks and Conservation Association	1919	42,000
Izaak Walton League	1922	60,000
The Wilderness Society	1935	68,000
National Wildlife Federation	1936	620,000
Nature Conservancy	1951	25,000
Defenders of Wildlife	1959	35,000
New Groups		
Environmental Defense Fund	1967	45,000
Friends of the Earth	1969	19,000
Natural Resources Defense Council	1970	35,000
Environmental Action	1970	16,000
Union of Concerned Scientists	1971	45,000
Cousteau Society	1975	150,000
New Directions	1976	11,000

Source: Mitchell, R. and Davies, C., III (1978) The United
States Environmental Movement and its Political Context: An
Overview, Discussion Paper D-32, Resources for the Future,
Washington, D.C., Table 2.

130

account when balancing environmental considerations
against other objectives are a notable advance.

Court litigation is not just a means to ensure
due process, it has also been exploited by US
environmental groups to question the policy underlying an agency action, e.g. by raising matters such
as a national nuclear energy policy and radioactive
waste disposal strategy when challenging an EIS for
a site licence for a nuclear power plant. This is
technically an improper move since the role of
court review is essentially to ensure that administrative procedures are properly followed. Recently
the Supreme Court ruled that the courts were not to
be employed as another forum for debating policy -
that was the job of Congress and the state legislatures (see O'Riordan, 1979b). This is a controversial judgement which has not been taken too
literally by the environmental groups who rightly
argue that the line between procedure and substance
is a fine one. Nevertheless they accept that litigation is an expensive and divisive mechanism and
its role in the early 1980s is bound to be more constrained and more carefully considered than was the
case a decade ago.

In Britain the courts play an almost negligible
role in environmental policy-making and execution.
Constitutionally they are subservient to Parliament
so can only interpret its will but not comment upon
the merits of legislation. Only when a minister or
his department acts unreasonably, capriciously or
contrary to established law can the courts intervene.
Otherwise citizen action is confined to legal intervention at common law involving individual cases of
nuisance, negligence or trespass. The use of
common law (law made by the courts themselves based
on the merits of a case and precedent drawn from
similar cases) is important in protecting basic individual rights to a reasonable degree of environmental
amenity but it normally only applies to persons and
to property owners and this cannot readily be
exploited by public interest environmental groups
intent on furthering a wider cause. One area where
citizen litigation in pollution control may be
furthered lies in the provision of the Control of
Pollution Act of 1974 which requires pollution control authorities to provide public registers of air
and water emissions (and noise emissions in certain
areas) and the standards (consents) which these
discharges are expected to meet. In theory this
should allow individuals or groups to take legal
action against a polluter where practice contravenes

licence.   However, it is unlikely that many pro-
secutions will ensue as very strict requirements
must be followed with respect to sampling before a
legally acceptable case can be heard, and in any case
the licences may be varied by the pollution control
authorities possibly to reflect the 'realities' of
the actual discharges (McRory and Zaba, 1978).
This is a sore point among environmental groups and
illustrates further the containment thesis advanced
earlier.   (In any case, this particular requirement
has not yet been put into practice.)

Implementation of Environmental Policy

Implementation of environmental policy takes place
through regulation and enforcement and the execution
of decisions backed up by budgets of authoritative
political bodies.   In Britain the executive
agencies pride themselves with a number of
attributes which they do not see paralleled elsewhere in the world.   These attributes are honesty
and incorruptibility, professional integrity and
professional ethos of competence, a belief in collaboration and co-operation with clients, a willingness to interpret the laws with discretion and
flexibility to suit circumstances, and a preparedness
to use enforcement powers if necessary as a last
resort.   Non-British regulatory officials are
astonished at these claims arguing that laws should
be more tightly laid down, that the concepts of
reasonableness and best practicable means should be
more clearly defined and that professional integrity
is surely subject to corrosion and manipulation from
time to time.   This is especially true of continen-
tal European environmental regulators who are sometimes asserted to be more susceptible to corruption
and political influence than their British counterparts and who depend far more explicitly on a
clearly enforceable set of regulations.   American
regulators are less corruptible but rely extensively
upon quite precise regulations, which they establish
on their own, often with little consultation with
their clients, and which they expect to be followed,
irrespective at times of the costs and technological
difficulties involved.

     For example, the US approach on nuclear safety
regulation is for the Nuclear Regulatory Commission
to prepare a set of detailed regulations on all conceivable aspects of safety as a result of exhaustive
discussions among its own staff (see Figure 4.4).

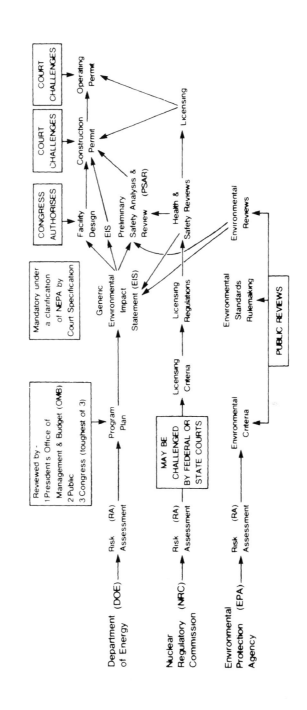

Figure 4.4: Main Procedural Steps Leading to Licensing of a Nuclear Reactor or for Safe Waste Disposal in the USA

This is done without any formal consultation either
with the major electrical utilities or with the
principal anti-nuclear groups who have their own
experts.   Eventually the rules and regulations are
ready for hearings at which various parties appear,
armed with their own specialist scientific and
technical briefs and accompanied by lawyers who
translate the arguments into legal language.   How
the final set of rules eventually emerges depends on
a host of variables such as the countervailing skills
of legal and technical experts, the mood of the
public, the interest taken by the politicians and
whether a nuclear 'incident' has occurred in the
recent past.   The points to grasp here are: (a) the
commitment to detailed documentation; (b) the
exhaustive justification of all arguments relating
to particular standards of assessment; (c) the
fascination with precise numerical indices; and (d)
the complex interaction of science, administration,
the law and politics in which economic analysis via
cost-effectiveness principles is not very prominent.
Implementation then follows the rules: the room for
manoeuvre between plant licensee (the utility) and
the regulator is fairly small and if rules are
broken the possibilities for prosecution either by
the regulatory agency or the public interest environmental groups are enormous.
     In Britain nuclear safety regulations are based
on a series of general guidelines laid down by the
Nuclear Installations Inspectorate following
elaborate consultation with the nuclear industry
(Figure 4.5).   These guidelines form the basis for
more detailed principles which are prepared not by
the regulatory agency but by the electricity
generating board (the utility) in collaboration with
the nuclear construction consortia (the manufacturers
and suppliers).   The boards are therefore responsible for preparing the safety case and for ensuring
that the plants are safe when in operation.   The
regulatory agency remains in the background throughout both phases of safety analysis and review but is
always on hand for advice when requested.   Thus the
British approach places the responsibility for safety
not on the regulator but on the operator, and the
standards of safety are also set by the operators
with the help of the regulator who provides general
guidelines.   Cost effectiveness of all safety
devices is an important item in the safety case and
regulating officials must take cost considerations
into account in their test of 'reasonableness'.
Another important difference between the US and UK

Figure 4.5: Risk Assessment and Decision Pathway for the PWR Proposed for Sizewell in Suffolk

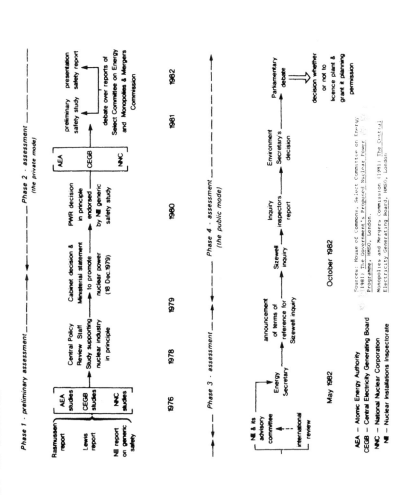

approaches lies in the flexibility with which the
rules are followed for Britain relies on self-
regulation by the industry as a prime feature of its
regulatory approach.   Finally consultation amongst
all interested parties takes place throughout the
whole exercise - but it is a contained consultation
for outside interests are not privy to these discussions until these are at an advanced stage and
clear positions have been established.   Hence the
strident calls for more openness and accountability
within British regulatory practice with respect to
environmental protection.
     Implementation of environmental policy takes
place through agency, civil service and local
authority officials who view their role differently
according to national context.   In Britain these
officials have a considerable degree of de facto
power - they act as gatekeepers to the public
interest - but would be most surprised and a little
affronted if told that this were the case.   They
would argue that they respond to internationally
and nationally accepted guidelines relating to
public health and safety and that they depend on
political authority and sanction to legitimate their
work.   In practice this political sanction is
rarely explicitly given so long as nothing goes
wrong.   It is when the public become aware of mis-
chief or mishap that the trouble occurs and the
politicians intervene.   Sometimes this means a
massive reinterpretation of practice - as for
example in the case of the control of toxic waste
dumps in the UK after children were discovered
playing with old cyanide canisters on an unregistered
waste tip near their homes - but usually an internal
inquiry is held and modifications are made about
which the public know little.   In the US the
officials play a much more prominent political and
public role for they are far more obviously accountable both to politicians and to the courts.   This
is why they shelter behind regulations which are
often unenforceable in practice and which are sometimes economically wasteful to implement in full.
     An interesting issue in the study of environ-
mental policy implementation is the extent to which
officials control the actual degree of flexibility
with which quangos and regulatory bodies have room
to manoeuvre to guide political commitment.   There
is no simple answer here for much depends on the
case in question and the prevailing public mood.
Certainly officials often 'lead from within'.   In
environmental matters especially elected members are

usually too busy and too ill informed and sometimes too trusting of their advisers to be able to counter-act official advice. Where they act they do so largely in response to political judgement (and here they are assisted by political advisers). Thus the ideology of officials with respect to environmental protection and development is an important element in understanding a national environmental strategy, such as it is. In the UK there is much discussion amongst politicians and political scientists as to the precise role played by civil servants in formu-lating and executing policy (see Mackintosh, 1974, and Sedgemore, 1980). Civil servants operate in accord with a code of collective self protection which inhibits them from admitting to the nature of their task, but many inside observers feel that they are very influential in guiding ministerial decisions toward acceptable policy based on what they regard as acceptable grounds for action. One should be careful in accusing civil servants of running a country, however. Their job is a powerful one but it is never dominant: political forces and judge-ments do count for much.

Executive agencies likewise play an important role where they can create a good case and capitalise on the mood of the times. The US Environmental Protection Agency has done this effectively with respect to the control of toxic chemicals while the UK Health and Safety Executive has been equally successful in its regulation of hazardous installa-tions. Here the gatekeeper not only opens the way to implementation but may provide the avenue leading to the entrance.

To study the implementation of environmental policy in practical terms is often a frustrating exercise for the interested student of public administration. The interaction of politics, economics, administration and the law make it all but impossible to disentangle all the relevant factors bearing upon a decision or leading to a particular policy. The key to success is to be aware of the basic principles some of which have been outlined in this chapter and to look carefully at case studies preferably from the perspective of an inside actor. This is by far the best way to get behind the walls of containment and to sense the subtleties of argument and positional movement as an issue is transferred from problem to decision.

Timothy O'Riordan

References

Anderson, F.R. (1973) NEPA and the Courts, Johns
      Hopkins University Press, Baltimore
Armitage, Sir A. (1980) Report of the Inquiry into
      Lorries, People and the Environment, Her
      Majesty's Stationery Office, London
Bachrach, P. and Baratz, M.S. (1962) 'Two Faces of
      Power', American Political Science Review, 56,
      947-52
Crenson, M.A. (1971) The Un-Politics of Air
      Pollution, Johns Hopkins University Press,
      Baltimore
Dahl, R.A. (1961) Who Governs? Yale University Press,
      New Haven, Connecticut
Drew, E.B. (1970) 'Dam Outrage: The Story of the
      Army Engineers', Atlantic Monthly, 225, 57-62
Dreyfus, D.A. and Ingram, H.M. (1976) 'The National
      Environmental Policy Act: A View of Intent and
      Practice', Natural Resources Journal, 16, 243-
      62
Fairfax, S.K. (1978) 'A Disaster in the Environmental
      Movement', Science, 199, 743-5
Friends of the Earth (1981) The Pressurised Water
      Reactor: A Critique of the Government's Nuclear
      Power Programme, Friends of the Earth, London
House of Commons, Select Committee on Energy (1981)
      The Government's Proposed Nuclear Power
      Programme, Her Majesty's Stationery Office,
      London
House of Lords, Select Committee on the European
      Communities (1981a) EEC Environmental Policy,
      House of Lords Paper No. 40, Her Majesty's
      Stationery Office, London
House of Lords, Select Committee on the European
      Communities (1981b) Environmental Assessment of
      Projects, House of Lords Paper No. 69, Her
      Majesty's Stationery Office, London
Ingram, H.M. (ed.) (1976) 'Agency Response to NEPA',
      Natural Resources Journal, 16, 240-384
International Union for the Conservation of Nature
      (1980) World Conservation Strategy, IUNC,
      Geneva
Jenkins, W.I. (1978) Policy Analysis, Martin
      Robertson, Oxford
Kasperson, R.E. (1980) 'US Nuclear Risk Management',
      unpublished paper, Center for Technology,
      Environment and Development, Clark University,
      Worcester, Massachusetts
Mackintosh, J.P. (1974) The Government and Politics

of Britain, Hutchinson, London

McRory, R. and Zaba, B. (1978) Polluters Pay: The Control of Pollution Act Explained, Friends of the Earth, London

Mitchell, R. and Davies, C., III (1978) The United States Environmental Movement and its Political Context: An Overview, Discussion Paper D-32, Resources for the Future, Washington, DC

Monopolies and Mergers Commission (1981) The Central Electricity Generating Board, Her Majesty's Stationery Office, London

O'Riordan, T. (1979a) 'The Role of Environmental Quality Objectives in the Politics of Pollution Control', in T. O'Riordan and R.C. d'Arge (eds.), Progress in Resource Management and Environmental Planning, Volume 1, Wiley, Chichester, pp. 221-58

O'Riordan, T. (1979b) 'Public Interest Environmental Groups in the United States and Britain', Journal of American Studies, 13, 409-38

O'Riordan, T. and Sewell, W.R.D. (1981) Project Appraisal and Policy Review, Wiley, Chichester

Outer Circle Policy Unit (1979) The Big Public Inquiry, London

Royal Commission on Environmental Pollution (1976a) Air Pollution: An Integrated Approach, Fifth Report, Cmnd.6371, Her Majesty's Stationery Office, London

Royal Commission on Environmental Pollution (1976b) Nuclear Power and the Environment, Sixth Report, Cmnd.6618, Her Majesty's Stationery Office, London

Scarman, L. (1974) English Law: The New Dimension, Stevens, London

Sedgemore, B. (1980) The Secret Constitution: An Analysis of the Political Establishment, Hodder and Stoughton, London

Shoard, M. (1980) The Theft of the Countryside, Temple Smith, London

US Council on Environmental Quality (1980) Global 2000, US Government Printing Office, Washington, DC

Wengert, N. (1955) Natural Resources and the Political Struggle, Doubleday, New York

Wynne, B.E. (1980) 'Windscale: A Case History in the Political Art of Muddling Through' in T. O'Riordan and R.K. Turner (eds.), Progress in Resource Management and Environmental Planning, Volume 2, Wiley, Chichester, pp. 165-204

Chapter 5

ENVIRONMENTAL POLICY AND INDUSTRIAL LOCATION IN THE UNITED STATES

Keith Chapman

Environmentalism has been defined as a code of conduct which involves the recognition of man's obligation to all other living and inanimate things (O'Riordan, 1976). The strength of the environmental movement has been acknowledged in most countries of the developed world during the last decade by the creation of institutional structures designed to ensure the incorporation of such values within the decision-making processes of both public and private agencies. In view of the important role of industry in economic growth, it is not surprising that much legislation inspired by environmental considerations should affect the activities of the manufacturing sector. Numerous studies have considered the economic implications of environmental policies for particular industries (see Atkins and Lowe, 1977; Thompson, Calloway and Nawalanic, 1978), but few have been concerned with the way in which these policies may influence the location of investment, either directly, through limitations upon the development of greenfield sites, or indirectly, as a result of differential rates of growth at existing sites (Chapman, 1980a). The balance of the literature reflects the fact that, apart from land-use controls and certain air pollution control measures, the various instruments of environmental policy are not specifically designed to influence the location of economic activities in the same way that regional policy, for example, has an explicitly spatial focus. Nevertheless there is evidence that environmental policies may have subtle and unexpected influences upon the location of manufacturing industry. The fact that these influences may be neither obvious nor consistent with other policy objectives makes them all the more worthy of consideration.

The necessity to conform with environmental
regulations may influence decisions relating to the
location of new manufacturing investment at a variety
of scales.   It has been suggested that the
existence of significant differences in environmental standards between developing and developed
countries may encourage a shift of polluting
industries towards the Third World (see Walter,
1976).   However, this argument assumes a considerable degree of locational flexibility at the macro-
scale and there is little evidence of such a trend
(see Gladwin and Welles, 1976; Pearson and Pryor,
1978; OECD, 1979).   At the other extreme, there is
no doubt that planning restrictions and environmental opposition have resulted in local, but significant, readjustments in siting.   For example, the
St. Fergus terminal, which is the major United
Kingdom landfall for gas from the North Sea, was
diverted to its present location from a site several
kilometers further north as a result of objections
to the routing of pipelines through an important
wild-fowl area.   However, it is at the inter-
regional scale that environmental policies may be
expected to have the greatest influence upon industrial location.   It is important to stress that
evidence of such an influence is limited and the
relatively recent introduction of more demanding
regulations makes it difficult to isolate the real,
as opposed to the alleged, impacts upon the
activities of manufacturing industry.   Nevertheless,
environmental regulations may be regarded as just
one element of the complex system of institutional
constraints imposed by governments upon corporate
behaviour.   This particular aspect of government
intervention is most apparent in the United States
which is, therefore, the best place to seek evidence
of environmental influences upon industrial location.
     Even in the United States, which has created
the most comprehensive package of environmental
legislation, official concern with such matters is
relatively recent.   Many states enacted legislation
at a very early date relating to, for example, the
pollution of streams and rivers, but such statutes
were usually regarded as legal curiosities rather
than effective policy instruments and it was only
during the 1960s that any significant measures were
introduced.   The political influence of the environmental movement is greater in certain states than in
others and this is reflected in the timing and
stringency of state legislation.   Inter-state
variations in land-use controls, for example, are

considerable (Rosenbaum, 1976). Recognition of
such variations and the opportunities which they
offer for prospective developers to minimise regulation of their activities by playing one state off
against another was an important factor encouraging
greater federal involvement in environmental matters
and eliminating these opportunities 'has been one of
the objectives behind federal intervention in air
and water pollution policy from the very beginning'
(Freeman, 1978, p. 43). Despite these efforts to
establish a more uniform environmental policy
climate during the 1970s, the nature and implementation of federal policies have ensured that they have
been by no means spatially neutral in their effects.
    Although the sheer bulk of Title 40 of the Code
of Federal Regulations (i.e. Protection of
Environment) emphasises the wide range of measures
introduced since 1970, it is convenient to distinguish between those aimed at encouraging the incorporation of environmental considerations within the
decision-making processes of potential developers
and those geared towards the control of specific
types of pollution. This distinction is somewhat
artificial, but it provides a convenient framework
within which different types of influence upon
industrial location may be considered. Accordingly,
the following chapter reviews the implications for
the location of manufacturing industry in the United
States of legislation relating to

    (i)   the preparation of environmental impact
           statements
    (ii)   the control of water pollution
    (iii) the control of air pollution.

Preparation of Environmental Impact Statements

The National Environmental Policy Act (NEPA) of 1969
represented the first major effort to encourage the
incorporation of environmental considerations within
the decision-making processes of federal agencies.
It also served as a catalyst to a succession of
related measures designed to implement the philosophy
outlined in the preamble to this piece of legislation which seeks '... to create and maintain conditions under which man and nature can exist in productive harmony and fulfil the social, economic and
other requirements of present and future generations'
(US Congress, 1969, Section 101(a)). The most
important single mechanism conceived by NEPA and

directed towards the attainment of these idealistic
objectives was the requirement to prepare an environmental impact statement (EIS) for any 'major
federal actions significantly affecting the quality
of the human environment'. An EIS is a document
containing the results of a systematic appraisal of
the potential impacts upon the environment of a
proposed development. Despite the apparently
narrow concern with 'federal actions' several
factors have ensured that the number of situations
in which an EIS is required have steadily increased
through time.

The courts have played the dominant role, not
only in the interpretation of NEPA, but also in
making it such a powerful influence upon the
American political scene over the last ten years
(see Anderson, 1973; Curlin, 1976). Thus the full
implications of each of the key words or phrases in
NEPA have been and are being established by legal
precedent. The overall result of this process has
been the adoption of a much more comprehensive
definition of the type of federal actions initiating
an EIS than was envisaged by the original proponents
of NEPA. Thus EISs are required not only for
projects undertaken directly by federal agencies,
such as flood control schemes carried out by the
Corps of Engineers, but also for projects in
receipt of funding from the federal government.
Private developments may also be incorporated within
the net as a result of the fact that the issue of a
license, for example a waste discharge permit under
the terms of the 1972 Water Pollution Control
Amendments, may also be construed as 'federal action'.

NEPA has inspired equivalent legislation at
state and county level and eighteen states had at
least some requirement for EIS preparation by 1979.
However, there are considerable variations in these
requirements associated not only with the definition of areas and developments for which EISs must
be produced, but also with the administrative responsibility for their preparation and review (Hart
and Enk, 1980). Superimposed upon these inter-
state variations are intra-state differences arising
from initiatives taken by local county administrations. Such differences may obviously encourage
business executives to distinguish between locations
in terms of the probable stringency of their EIS
requirements. However, the substance of these
requirements is probably less significant than their
mere existence in influencing corporate images of
place and, although an EIS is by no means an

automatic consequence of a major development proposal in the United States, the trend towards a
more widespread adoption of this practice has had
some important consequences for decision-making in
industry.

By providing a focus for organised opposition,
the EIS procedure has tended to make the development
of new sites more difficult. For example, doubts
regarding the accuracy and objectivity of the EIS
were an important factor in the delays which prompted
Dow Chemical to abandon in 1977 a plan to construct
a major petrochemical complex at the mouth of the
Sacramento River in California (Storper, Walker and
Widess, 1981). In this context, it has been
observed that the EIS requirement created additional
uncertainties from Dow's point of view (Brubaker,
1978). Thus, viewed from a corporate standpoint,
not only do EIS requirements raise the possibility
that a development may be prevented altogether, but
they also tend to result in delays which increase
the project planning period.

Increased public involvement is not the only
contributor to additional delay and uncertainty
arising from environmental considerations. This
trend is further reinforced by the need to comply
with more pervasive and demanding regulations.
Elkin and Constable suggest that 'It would be a
conservative estimate to say that the 1977 Air Act
Amendments add 1 to 3 years to the planning and
development of a new or modified major (industrial)
source locating anywhere in the United States'
(1978, p. 296). Despite the co-ordinating role of
the Environmental Protection Agency (EPA), there is
no doubt that the proliferation of federal legislation governing the environment has been accompanied
by a fragmentation of agency responsibility which
has been a further source of delay in, for example,
the establishment of new power stations (Landsberg,
1979, pp. 519-521). Yet more opportunities for
procrastination are implicit in the federal
political structure of the United States as disputes
have arisen between Washington and the state
capitals over the interpretation and enforcement of
environmental legislation. Although the nature of
the regulatory system is such that these various
sources of uncertainty may apply to major investments at existing as well as greenfield sites, the
problems facing the would-be developer are certainly
more acute in a new location and environmental
policy considerations seem likely to reinforce the
already firmly established preference for in situ

expansions.

An EIS may be regarded by companies as an unwelcome legislative hurdle to be overcome before
many industrial developments can proceed. However,
these external pressures may be expected to induce
more careful internal assessments of environmental
issues at the project planning stage.   It would be
naïve to expect private (and public) enterprises to
make the maintenance of environmental quality a
primary goal, but government may be able to create a
situation in which the incorporation of such considerations within the decision-making processes is
consistent with corporate self-interest (Rothenberg,
1974).   Despite Gladwin's pessimistic conclusion
that 'Natural processes are often disregarded and
full consideration of environmental values is
typically absent' (1975, p. 256) in the project
planning behaviour of multinational corporations,
there is evidence of some change in this situation.
Royston (1979) emphasises the positive contribution
which pollution control may make to corporate
balance sheets whilst, in the United Kingdom,
executives of both British Petroleum (1979), which
operates as a multinational corporation, and the
nationalised British Gas Corporation (Dean and
Graham, 1978) justify the use of environmental
impact assessments within their own organisations
on the grounds that costly delays in gaining the
necessary permits and approvals are reduced as a
result of the avoidance of conflicts with environmental groups.

However desirable this kind of homework may be,
it is generally carried out after the location
decision has been made and examples of cases in
which internal corporate assessments of environmental impacts have actually resulted in the rejection
of a site previously selected on other grounds are
probably extremely rare.   Indeed evidence of such
a situation could only be assembled in the unlikely
event of independent access to the decision-making
procedures of individual companies.   In the absence
of studies utilising this approach, intuitive
judgement suggests that legislation relating to EIS
preparation can only be regarded as a minor influence upon plant location decisions.   By contrast,
the definition of certain thresholds in connection
with the control of air and water pollution has
created circumstances in which environmental factors
may assume much greater significance as influences
upon the spatial distribution of manufacturing
investment.

Control of Water Pollution

Despite the preference of most academic economists
for the use of tax mechanisms, policies aimed at
limiting pollution from industrial plants generally
rest upon the definition of standards which may be
expressed either in terms of the quality of the
receiving environment or the properties of the discharges themselves.   The adoption of one or other
of such ambient and point source standards may have
significant implications for manufacturing industry
as changes in the direction of United States' water
pollution control policy during the last fifteen
years clearly demonstrate.   Before 1972, the starting point in the regulation of water pollution from
industrial plants was the definition of ambient
water quality standards.   Initially these were
determined by the appropriate state agencies, but
the Water Quality Act of 1965 established federal
standards for inter-state waters.   The quality and
quantity of permitted discharges from individual
polluters on a particular water course were derived
by estimating the capacity of the receiving stream
to assimilate effluent and still meet the specified
ambient standard.   Implicit in this approach is
the possibility that spatial variations in the
assimilative capacity of the environment may become
a variable in the location decision because of their
impact upon pollution control costs.   However, the
widespread reluctance to enforce the pre-1972
legislation meant that this possibility remained of
theoretical rather than practical significance.
     Evidence that the problem of water pollution
was increasing despite the introduction of ambient
water quality standards (see Kneese and Schultze,
1975, pp. 42-45), prompted a radical reappraisal
of policy which was embodied in the 1972 Water
Pollution Control Amendments.   By establishing the
admittedly unrealistic goal of eliminating by 1985
all discharge of pollutants into navigable waters
regardless of cost this legislation rejected the
notion that a certain economically optimal level of
pollution is acceptable.   An interim objective was
also declared of achieving by 1983 'water that is
clean enough for swimming, boating and protection
of fish, shellfish and wildlife' (US   EPA, 1976,
p. 2).   This latter objective was to be achieved
by improving specific standards on particular
categories of point source rather than by writing
permits on a case by case basis depending upon a

presumed relationship between discharges and water
quality. The complex and tenuous nature of this
relationship had been a problem with the 1965
legislation and it had been argued that permit conditions defined in this way were almost certainly unenforceable in law (Zurick, 1971).   Quite apart
from its legal implications, this change apparently
removed the scope for selecting locations on the
basis of differences in required levels of pollution
control.

A succession of technical documents has been
produced by or on behalf of the EPA with the objective of establishing acceptable standards of
effluent from a wide range of industrial plant types
(see US EPA, 1974).   These documents, usually prepared by engineers and subsequently enforced by
lawyers, have resulted in a proliferation of jargon
such as Best Practicable Technology (BPT) and Best
Available Technology (BAT).   Although intended to
define unambiguous standards, these terminologies
continue to provide opportunities for spatial
variations in pollution control requirements.
Despite the fact that ultimate responsibility for
validating the conditions laid down in discharge
permits lies with the EPA, there is evidence that
inter-state differences remain.   For example, an
analysis of the effluent characteristics of a
number of petrochemical plants in Texas and
Louisiana over the period 1970 to 1978 suggests
that, despite an overall improvement in quality in
both states, standards of treatment were generally
higher in the former than the latter   (Chapman,
1980b).   These differences may be tentatively interpreted as a consequence of the adoption of more
vigorous pollution control policies by the appropriate state agencies in Texas as compared with their
counterparts in Louisiana.   The same study also
revealed intra-state differences related partly to
political factors and partly to the continued recognition of assimilative capacity as a relevant
variable in defining what quantities and qualities
of effluent discharge are acceptable.   In Texas,
Harris County has acquired a reputation for
relatively vigorous enforcement of pollution control
statutes and it has been argued that this local
initiative has focused the attention of state and
federal agencies upon the problems posed by the
massive concentration of oil refining and petrochemical capacity along the length of the Houston
Ship Channel (Powelson, 1978).   Thus officials of
the Texas Department of Water Resources insist that

the conditions imposed upon dischargers to this waterway are more demanding than the BPT require- ments laid down in the effluent guideline documents of the EPA. This suggests that dischargers to the Ship Channel may be faced with the need to install more costly water treatment systems than similar plants located elsewhere in Texas.

Politics play an important role in the imple- mentation and enforcement of pollution control policy. Although the trend towards uniform techno- logy-based effluent standards for particular types of industrial process is narrowing the areas of un- certainty, the conditions imposed in waste dis- charge permits are normally the outcome of lengthy discussions between the issuing agency and the applicant company. Personalities may be important and the terms of a discharge permit may vary depend- ing upon its author, with certain senior staff within the same agency adopting a more lenient interpretation than others of what constitutes BAT. Major companies with large and complex manufacturing sites are often in a stronger negotiating position than smaller enterprises, especially where a single operation dominates a local economy. It should be stressed that not all larger companies exploit the political advantages conferred by their size and economic importance, but there is no doubt that it is often easier to impose and enforce strict standards upon small rather than large dischargers. Indeed, this is one of the reasons for greater federal intervention through the operation of the National Pollution Discharge Elimination System (NPDES) introduced by the 1972 Water Pollution Control Amendments. The opportunities for the exercise of political influence should theoretically be reduced with the transfer of ultimate responsibi- lity for permit conditions from the state agency to the EPA. For example, correspondence between Dow Chemical and the EPA relating to the terms of permits for the massive Freeport, Texas petro- chemical complex applied under the NPDES system suggests that the company has, at various times, regarded the federal agency as less reasonable by comparison with its previous experience of dealing only with the appropriate state authorities (Chapman, 1980b).

In addition to what may be loosely termed political advantages, size also confers certain technical and economic benefits. There is ample evidence of significant economies of scale in the technology of pollution control and a desired

quality of effluent from a particular industrial
process may be achieved at lower unit cost for a
large than for a small plant (see Russell and
Spofford, 1973; Leone, 1976). One response to
this situation has been the establishment of
regional treatment systems. This not only enables
smaller companies to enjoy the economic benefits of
participating in a large-scale operation, but also
to discharge some of their pollution control responsibilities to specialist firms. Technology-
forcing legislation is especially hard on small
enterprises which frequently lack the expertise to
solve their own pollution problems and the financial
resources to absorb the costly mistakes which are
inevitably made in such circumstances. Regional
treatment systems have generally been constructed
to meet the needs of existing concentrations of
industry, but they may also be established as an
integral part of new manufacturing complexes. For
example, one of the most important single factors
encouraging the concentration during the late 1970s
of several chemical firms at Bayport, approximately
35 km southeast of Houston, has been the construction of a joint-user waste treatment facility.
Although such facilities may promote agglomeration,
the technology of waste water treatment may also
have the reverse effect of encouraging dispersal.
Several observers have noted that shortage of space
may prevent the installation of pollution control
equipment on congested urban sites and therefore
contribute to the suburbanisation of manufacturing
(see Atkins and Lowe, 1977; Stafford, 1977). Even
where relocation does not occur, significant
savings in pollution control costs may be gained by
firms with large sites. For example, the option
of using extensive pond systems, which are land-
rather than capital-intensive, may be available to
companies in isolated locations whilst competitors
with more restrictive sites may be forced to employ
more costly technical solutions.
    Water pollution control policy in the United
States has, since 1972, rested upon the premise that
the various categories of discharger should attain
common standards of treatment regardless of
location. However, it is apparent that, in practice, this objective is very difficult to achieve.
Nevertheless, the 1972 Water Pollution Control
Amendments have created a national system which has
largely eliminated the possibility that inter-state
variations in policies relating to the protection of
streams and rivers could ever become a significant

variable in industrial location decisions within
the United States. By contrast, certain aspects
of federal legislation relating to the control of
air pollution are specifically designed to influence
the spatial distribution of economic activity.

Control of Air Pollution

Much of the impetus to air pollution control
legislation in the United States has been provided
by the problem of automobile emissions. Federal
measures in 1955, 1963 and 1965 focused upon this
question and little attention was devoted to emissions from industrial sources. The 1967 Air Quality
Act followed the pattern set two years earlier by
the Water Quality Act by requiring states to
establish ambient air quality standards on specified
pollutants such as sulphur oxides and particulates,
together with implementation plans for ensuring
their attainment. It was envisaged that these
plans would necessitate restrictions upon emissions
from industrial sources of pollution and that the
severity of these restrictions would vary from
place to place, depending upon prevailing air
quality conditions. In fact, the results of the
1967 Air Quality Act were even more disappointing
than those of the corresponding water pollution
control legislation. By 1970 not a single state
had a full-scale plan of standards and implementation in effect, and it was estimated that the
process would not be completed until well into the
1980s (Kneese and Schultze, 1975, p. 50).
 The 1970 Clean Air Amendments, which transferred
the responsibility for defining ambient air quality
standards to federal level, were a response to this
problem. As a direct result of the Amendments, the
EPA devised standards for a series of 'criteria'
pollutants including carbon monoxide, nitrogen
dioxide and suspended particulates. The mechanisms for achieving these standards were to be outlined in implementation plans prepared initially by
the states, but ultimately requiring approval from
the EPA. Furthermore, the EPA was empowered to
replace all or part of an implementation plan that
it considered inadequate with regulations that were
binding on the state. Although the 1970 legislation sharply increased limitations on motor vehicle
emissions, control of pollution from industrial
sources was also regarded as an integral part of
these plans. Existing and new plants were to be

treated differently.   In the former case, the
emission limitations specified in individual permits
were to be derived in the manner outlined in the
1967 Air Quality Act - by assuming a relationship
within an area between emissions and air quality and
then allocating the total acceptable discharge of
pollutants between individual plants.   A paradoxical
consequence of this somewhat pragmatic and scienti-
fically dubious approach is that, by requiring the
attainment of uniform national air quality goals,
the 1970 legislation has raised the possibility of
significant spatial variations in pollution control
costs for individual dischargers depending upon the
relationship between prevailing environmental conditions and the federal standards, as will be discussed below.   This possibility was theoretically
eliminated in the case of new plants which were
obliged to meet common standards of performance
defined by the EPA on the basis of technical studies.
These standards were independent of location and,
like the effluent restrictions embodied in the 1972
Water Pollution Control Amendments, abandoned the
notion of directly linking residuals discharge to
ambient environmental conditions. Despite the intentions of the 1970 Clean Air Amendments and the
voluminous guidelines provided by the EPA, it is
almost impossible to ensure that new facilities
locating in different states will receive identical
treatment since permit conditions for major plants
are, in reality, resolved on the basis of lengthy
negotiations between the applicant company and the
appropriate state and federal agencies.
     It is widely acknowledged in the literature of
environmental economics that relocation of industrial plants is a legitimate objective of pollution
control policies in general (see Baumol, 1972;
Burrows, 1979) and air pollution control strategies
in particular (see Köhn, 1974; Shefer and Guldman,
1975; Tolley, 1975).   In preparing their implementation plans as required by the 1970 Clean Air
Amendments, most states anticipated a convergence of
air quality within their boundaries with improvements in 'dirty' areas and controlled growth in
'clean' areas provided that any consequent deterioration did not result in violation of the federal
ambient standards.   This strategy implied some re-
distribution of economic growth away from metropolitan centres in favour of more peripheral locations.
However, it is doubtful that the instigators of the
1970 Clean Air Amendments in the United States
fully appreciated the potential implications of their

actions for the spatial distribution of manufacturing (Smith, 1977). These implications became more
apparent as a result of a series of interpretive
rulings by the courts which established that air
quality should be prevented from deteriorating
'significantly', even in areas satisfying the
various ambient standards specified in the 1970
legislation (see Stern, 1977).

The definition of 'significantly' was obviously
critical and it became apparent that an extreme
view would impose a veto on any further heavy industrial development in any part of the United States
(see Freeman, 1978, p.40). Faced with this
prospect, the EPA has devised a more flexible
approach which has, nevertheless, been described as
'the most formidable package of land-use restraints
set forth by Congress' (Elkin and Constable, 1978,
p. 287). This package was enshrined in the 1977
Clean Air Amendments which effectively distinguish
between two general types of area - those in which
air quality meets the relevant federal ambient
standards and those which are designated non-
attainment areas. Restrictions upon pollution
sources vary depending upon their location with
respect to this geographical classification. The
nature of these restrictions is elaborated within
the terms of the so-called 'prevention of significant deterioration' (PSD) and 'offset' programmes
incorporated within the 1977 legislation. It
should be noted that the operation of these programmes is complicated by the fact that attainment
and non-attainment areas are not necessarily
mutually exclusive. The definitions apply to each
of the individual criteria pollutants and an area
may therefore fail to meet the federal standard for
one of the pollutants but at the same time be
regarded as satisfactory in terms of the others.

The PSD programme requires state agencies to
allocate all attainment areas within their jurisdiction to one of three categories according to the
anticipated level of control upon new pollution
sources. A requirement for 'absolute protection of
visibility' in Class I areas, which typically
include National Parks and Wilderness Areas, implies
virtually no deterioration in air quality.
Moderate deterioration of air quality, defined in
terms of specific incremental changes in ambient
concentrations, is permitted in Class II areas which
account for virtually all of the remaining
territory with attainment status. However, a state
governor has the authority to designate Class III

areas of 'intensive economic growth' in which larger
increments in ambient concentrations are permitted
up to the level of the national secondary standards.
Both primary and secondary national standards for
ambient concentrations of the various criteria
pollutants have been devised.   The former were
devised with reference to presumed adverse effects
upon health, the latter are more stringent and seek
to protect 'the public welfare from any known or
anticipated adverse effects'.

    By establishing three categories of attainment
area, the PSD programme formalises the preservation
of geographical differences in air quality which
may otherwise have been eroded if the state implementation plans spawned by the 1970 Clean Air Amendments had remained in force.   It also allows,
through the mechanism of class III areas, for some
balancing of economic benefits and environmental
costs, a judgement which was specifically excluded
by any literal interpretation of the 1970 legislation (Freeman, 1978, p. 40).   This in turn allows
the possibility of spatial variations in pollution
control requirements.   Such variations are
probably not significant enough to influence location decisions, but the PSD programme has created
serious obstacles to any future industrial development over large areas of the Western United States.
Class I status virtually eliminates the possibility
of introducing any major pollution source to an
area.   Indeed the requirement for 'absolute protection of visibility' in such areas may prevent the
establishment of, for example, a coal-burning power
station in an adjacent Class II area since fine
particulates from such a facility could affect
this attribute of air quality hundreds of kilometers from the source (Landsberg, 1979, p. 383).
Restrictions on raw material oriented industries
such as copper smelting could prevent the extraction of valuable resources altogether.   Nowhere is
this issue more apparent than in the debate over
the exploitation of the massive coal reserves of
such states as Montana, Wyoming and Colorado.   The
implications of the PSD policy for the mining and
subsequent utilisation of western coal is one of
the most striking of the many conflicts between
energy and environmental policy objectives in the
United States (see Goodwin, 1981).

    Whereas the PSD policy tends to have the
greatest impact upon relatively undeveloped areas,
the offset strategy is directed primarily at
existing centres of economic activity which are

designated as non-attainment areas with respect to
the federal ambient standards.   Further industrial
development in such areas is permitted only if a new
source can demonstrate that its increment to the
total volume of emissions will be more than offset
by compensating reductions elsewhere within a somewhat ill-defined proximal zone.   Any new plant
locating in a non-attainment area is also required
to install technology aimed at securing the Lowest
Achievable Emission Rate as defined by the EPA.
This may impose high pollution control costs upon a
new plant whereas 'For existing sources, there is
little more than the traditional regulatory
pressure to do one's best to reduce emissions if it
is not too costly.' (Landsberg, 1979, p. 396).
Despite this discrimination, the principal feature
of the offset policy is the introduction of what
has been labelled a 'market in dirty air' as a means
of evading the barrier to economic growth in non-
attainment areas imposed by the 1970 Clean Air
Amendments.
     Several methods of acquiring offsets have
emerged as state agencies and private industry have
attempted to interpret and implement federal policy.
Large existing pollution sources generally have
little difficulty in compensating for new facilities
on a major site by making reductions elsewhere on
the same site.   Alternatively, it may be possible
for a new source owner to finance the necessary
improvements in the existing facilities of another
company.   For example Sohio obtained permission to
construct an oil refinery at Long Beach, California
by agreeing to pay for control equipment in various
dry-cleaning establishments and electricity
generating plants.   Direct purchase of pollution
rights may occur, as when General Motors was
allowed to establish assembly plants in Shreveport,
Louisiana and Oklahoma City following agreements
with owners of nearby hydrocarbon storage facilities.
Yet another approach directly involves state
agencies in the bargaining process.   For example,
the Highway Department of the State of
Pennsylvania undertook 'to reduce its use of cut-back
asphalt for surfacing roads to enable the construction of a motor vehicle plant at Pittsburgh.   A
similar agreement was reached involving the corresponding agency in Texas to allow ICI Americas to
establish a petrochemical facility at Bayport near
Houston.
     Although the offset policy introduces a rudi-
mentary market mechanism of the type generally

favoured in the literature of environmental economics, the operation of this market is far from
perfect (McKee, 1978).   Since offsets are calculated
with reference to existing baselines, the policy may
penalise those states and counties which have
previously been most conscientious in pollution
control.   The same argument applies to companies,
with formerly lax dischargers having the advantage
of greater opportunities for securing future offsets at little cost. Viewed at a regional scale,
areas with a high proportion of heavily polluting
industries also have more scope for emission reduction than those in which these types of economic
activity are under-represented.   Thus   the
operation of the offset policy raises a number of
complex issues which are perhaps best understood
by considering some of the implications of the 1977
Clean Air Amendments for specific industries and
areas - in this case oil refining and petrochemicals
in Texas and Louisiana.
      Texas accounted for 27.4 per cent of United
States refining capacity in 1980, more than twice
the share of any other state, whilst the corresponding figure for Louisiana was 12.9 per cent (Pennwell,
1980, p. 405).   Their dominance of the petrochemical industry is even greater.   Using ethylene,
which is generally regarded as the most important
petrochemical intermediate, as a yardstick, 67.6 per
cent of total domestic capacity was located in
Texas in 1978 and 23.6 per cent in Louisiana (Wett,
1978).   This spatial concentration of petroleum-
related industries in a national context is duplica-
ted at the state level with virtually all of the
major complexes located in a belt paralleling the
Gulf Coast from Corpus Christi in the west to New
Orleans in the east.   Even within this zone, further
distinctive clusters may be identified in locations
such as the Houston Ship Channel, the 'Golden
Triangle' of Beaumont-Orange-Port Arthur, Lake
Charles and along the Mississippi between Baton
Rouge and New Orleans (Figure 5.1).   The characteristic agglomeration of these industries
accentuates the inevitable pollution problems and
makes it difficult to reconcile environmental policy
objectives with the continued growth of local and
regional economies which largely depend upon petroleum processing.
      As far as air pollution is concerned, oil
refineries and petrochemical plants are major
sources of hydrocarbons.   For example, almost 60
per cent of the total recorded emissions of

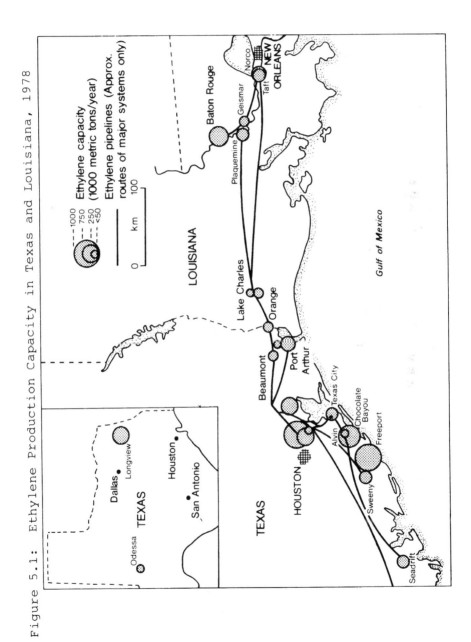

Figure 5.1: Ethylene Production Capacity in Texas and Louisiana, 1978

157

hydrocarbons in Texas in 1973 were derived from
such installations together with associated storage
facilities (Texas Air Control Board, n.d.).
Federal ambient standards for hydrocarbons were
originally introduced in 1971 because of the presumed role of these compounds in the formation of
photochemical oxidants.   These secondary pollutants,
more familiarly termed 'smog', are known to have
a wide range of toxic effects and are derived by a
complex series of chemical transformations based
upon oxides of nitrogen and hydrocarbons.   The
basic approach to the control of photochemical
oxidants adopted by the EPA has been to restrict
emissions of all non-methane hydrocarbons rather than
to focus attention upon nitrogen dioxide.   It is
argued that by limiting emissions of one of the
essential ingredients in the process, oxidant formation will be correspondingly reduced.   Whatever the
doubts regarding the scientific validity of this
policy (Tannahill, 1976), it has certainly directly
affected the activities of the oil refining and
petrochemical industries.

The primary ambient standard for photochemical
oxidants (measured as ozone) was set in 1971 at
0.08 ppm not to be exceeded for more than one hour
per year.   It soon became apparent that large
areas of the country had little prospect of attaining
this standard by the required dates.   More significant from the oil industry's point of view, it
was estimated that approximately 85 per cent of
United States oil refining and petrochemical
capacity in 1975 was located in photochemical
oxidant non-attainment areas (Radian, 1977).   A
similar pattern was apparent in Texas (Figure 5.2)
and Louisiana (Figure 5.3) with non-attainment
counties correlating fairly closely with the distribution of these industries.   This situation, with
its implications for the stringent control of hydrocarbon emissions, resulted in predictable reactions
from the petroleum industry.   Serious shortages in
refining capacity were projected at national level
(Elkin and Constable, 1978) and an industry representative at a hearing in Austin concerning the
introduction in 1972 of a state regulation to control hydrocarbon emissions suggested that it 'would
result in the shutdown of every refinery and petrochemical plant in Texas' (quoted in Magee and
Cooper, 1978, p.7).   Although the remark should be
viewed as a political statement rather than a
scientific prediction, it is symptomatic of oil
industry attitudes towards the 1970 Clean Air

Figure 5.2:   Photochemical Oxidant Non-attainment Counties in Texas

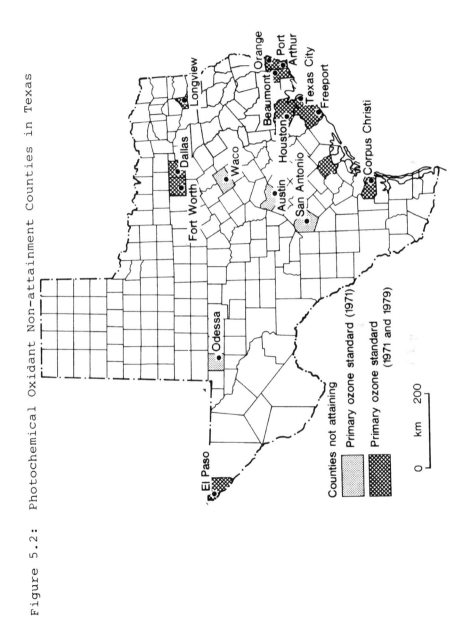

Amendments.   The offset mechanism introduced in
1977 represented an improvement from the industry's
viewpoint, but intensive lobbying by organisations
such as the American Petroleum Institute and municipalities such as the City of Houston to secure a
revision of the photochemical oxidant standard continued.   This activity contributed to the relaxa-
tion of the standard in 1979 to 0.12 ppm (US   EPA,
1979a).   In the case of Texas and Louisiana this
change did not affect the position of the oil
refining and petrochemical industries.   Only four
counties in Texas and none in Louisiana changed
their designation as a result of the modified
standard.   However, only Ector county (Odessa) of
the four Texas counties is associated with oil
refining and petrochemical activities and all of
the remainder which were designated non-attainment
with respect to the 1971 standard also fail to meet
the 1979 standard.   Thus the petroleum processing
industries of the Gulf Coast remain subject to the
restrictions associated with location in a non-
attainment area (Figures 5.2 and 5.3).
    These restrictions present few problems to
large existing sources of hydrocarbon emissions.
A detailed study of the opportunities for offsets
within the Houston-Galveston Air Quality Control
Region, which includes the non-attainment counties
of Harris, Galveston and Brazoria, concluded that
the future growth of established oil refining and
petrochemical complexes would not be prejudiced by
the offset policy (Engineering Science, 1978).
However, this policy is less easy to manipulate by
firms which are not already established in an area.
Such firms cannot secure the necessary offsets by
making improvements on existing sites or, as in the
case of a new petrochemical plant at Corpus Christi,
by taking similar measures in a nearby facility.
Generally speaking, it is naïve to expect that
established companies may be persuaded to reduce
their own emissions to enable a potential competitor
to set up a new plant and the 'market in dirty air'
is therefore unlikely to function very effectively.
Indeed, there is clear tactical advantage in the
inflation of emissions by continuing to operate
relatively old equipment as a 'bank' of future
offsets.   Thus in seeking its intended objective
of limiting the growth of emissions in non-
attainment areas, the offset policy is having the
unintended consequence of discriminating between
existing and new firms within such areas.
    Mounting evidence that photochemical oxidants

Figure 5.3: Photochemical Oxidant Non-attainment
Counties in Louisiana

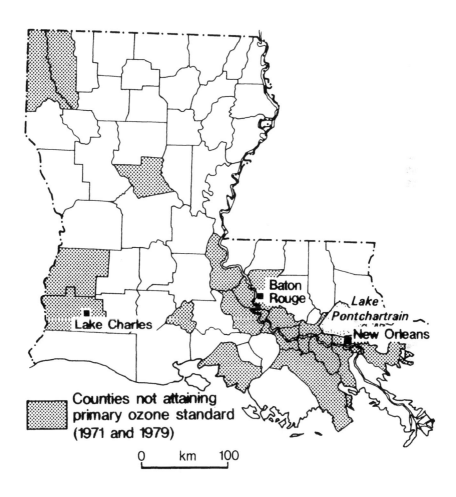

are not an exclusively urban problem and that concentrations of naturally occurring ozone in certain
rural areas approached or even exceeded the 1971
primary standard was a factor in its revision in
1979.   This evidence also resulted in a modification
of the offset policy which has tended to strengthen
its potential influence upon plant location
decisions (US   EPA, 1979b).   As far as the photochemical oxidant standard is concerned, the offset
policy will only apply in those non-attainment
counties where anthropogenic sources of hydrocarbons are assumed to be primarily responsible for
the high ozone levels.   For the purposes of implementation, the EPA has defined an arbitrary population level of 200,000 to distinguish between
'urban' counties, in which the problem is regarded
as man-made, and 'rural' counties, in which it is
viewed as a natural phenomenon.   This administrative device will have the effect, for example, of
making it easier to obtain the necessary permits
for any oil refining and petrochemical developments
in such 'rural' counties of Texas as Galveston and
Orange in comparison with their 'urban' neighbours
of Harris and Jefferson respectively.

Conclusions

It is difficult to provide any clear general statement regarding the influence of environmental policy
measures upon industrial location in the United
States.   This partly reflects the fact that there
must be a significant time-lag between the implementation of more demanding pollution control regulations and any observed effects upon plant location.
It is also due to the paucity of empirical work in
this area.   An attempt by Healy (1979) to relate
differential rates of employment growth in manufacturing in the various states of the United States
between 1970 and 1976 to spatial variations in the
severity of environmental regulations proved in-
conclusive.   Nevertheless, it would be surprising
to find that such a vague concept as Healy's
environmental regulation index provided any significant explanatory power at this scale.

A more direct approach involves simply asking
executives what importance they attach to environmental regulations in making location decisions.
However, it is difficult to isolate facts from
corporate rhetoric and such surveys are subject to
the same methodological problems of post hoc

rationalisation as any questionnaire-based study of
industrial location decisions.   Nevertheless
recent studies of this type do suggest that the
political influence and activity of environmental
pressure groups, for example, has a bearing upon
that intangible phenomenon known as business
climate (Stafford, 1979).   The true importance for
the geography of manufacturing of such executive
images of place is difficult to establish.
California is noted for its stringent pollution control requirements and the abandonment in 1977 of
plans for a major petrochemical complex in Solano
County 55 km north east of San Francisco, ostensibly
on the grounds of excessive government regulation
(Brubaker, 1978), is consistent with this image.
This case is probably typical of the way in which
environmental regulations affect location decisions
in the sense that the influence is negative rather
than positive; pushing facilities away from controversial sites rather than pulling them towards
places with less demanding standards.
       Although specifically concerned with the inter-
relationships between environmental policies and
industrial location in the United States, the
material assembled in this chapter has broader
implications for approaches to the study of the
geography of manufacturing.   Such policies are
being adopted with varying degrees of commitment by
many countries besides the United states (see OECD,
1979).   In the European Economic Community, for
example, there is ample evidence, in the form of
proposals relating to environmental impact assessments, air and water quality standards (Commission
of the European Communities, 1979), that the
Commission would like environmental policy in the
Community to follow much the same path as that
pioneered by the EPA in the United States.   These
proposals are symptomatic of a general trend towards
greater government involvement in the activities of
manufacturing industry in general and large enterprises in particular (see Watts, 1980, pp. 270-81).
With the exception of regional policy, the spatial
implications of such involvement are frequently
unanticipated. Close examination of the implementation of and responses to particular pieces of legislation serves to emphasise the growing role of the
state as a factor influencing the location of
economic activity and helps to isolate inconsis-
tencies between the actions and objectives of
different branches of government.   More specifically, the implementation of environmental policy

in the United States has a very significant
political dimension involving not only the relationships between government and private industry, but
also the division of responsibility between federal,
state and county authorities within the administrative hierarchy. Organisations such as the Texas
Air Control Board, for example, have little
enthusiasm for certain aspects of their role as
agents of federal authority and are not always in
sympathy with the regulations they are required to
enforce. These kinds of situation emphasise the
need for students of industrial location to
appreciate that investment decisions are taken within a complex legislative and political environment
and can rarely be understood in exclusively economic
terms.

References

Anderson, F.R. (1973) NEPA in the Courts: A Legal
    Analysis of the National Environmental Policy
    Act, Johns Hopkins University Press, Baltimore
Atkins, M.H. and Lowe, J.F. (1977) Pollution Control
    Costs in Industry, Pergamon Press, Oxford
Baumol, W.J. (1972) 'On Taxation and the Control of
    Externalities', American Economic Review, 62,
    307-22
British Petroleum (1979) 'BP Briefing Paper on
    Environmental Protection', Environmental Data
    Services, 20, 19-23
Brubaker, R.I. (1978) 'A Rocky Road Awaits Industry
    on its Way through the Environmental Obstacle
    Course', Industrial Development, 147, 2-5
Burrows, P. (1979) The Economic Theory of Pollution
    Control, Martin Robertson, Oxford
Chapman, K. (1980a) 'Environmental Policy and
    Industrial Location', Area, 12, 209-16
Chapman, K. (1980b) Petrochemicals and Pollution in
    Texas and Louisiana, Department of Geography,
    University of Aberdeen
Commission of the European Communities (1979) State
    of the Environment, Second Report, Commission
    of the European Communities, Brussels -
    Luxembourg
Curlin, J.W. (1976) 'The Role of the Courts in the
    Implementation of NEPA' in M. Blisset (ed.),
    Environmental Impact Assessment, Engineering
    Foundation, Austin, Texas, pp. 27-44
Dean, F.E. and Graham, G. (1978) 'The Application of

Environmental Impact Analysis in the British
Gas Industry (General Considerations)',
Proceedings of the UNECE Symposium on the Gas
Industry and the Environment, pp. 151-71
Elkin, H.F. and Constable, R.A. (1978) 'Impact of
the Non-attainment Policy of the Clean Air Act
on the Petroleum Industry', Proceedings,
American Petroleum Institute Summer Meeting
(Refining Group), pp. 287-300
Engineering Science (1978) An Analysis of Alterna-
tive Policies for Dealing with New Source
Growth in Non-attainment Areas, Volume 2:
Oxidants, Engineering Science Inc., McLean,
Virginia
Freeman, A.M., III (1978) 'Air and Water Pollution
Policy' in P.R. Portney (ed.), Current Issues
in U.S. Environmental Policy, Johns Hopkins
University Press, Baltimore, pp. 12-67
Gladwin, T.N. (1975) Environment, Planning and the
Multinational Corporation, Jai Press, Greenwich,
Connecticut
Gladwin, T.N. and Welles, J.G. (1976) 'Environmen-
tal Policy and Multinational Corporate
Strategy' in I. Walter (ed.), Studies in
International Environmental Economics, Wiley,
New York, pp. 119-224
Goodwin, C.D. (ed.) (1981) Energy Policy in
Perspective: Today's Problems, Yesterday's
Solutions, Brookings Institution, Washington,
DC
Hart, S.L. and Enk, G.A. (1980) Green Goals and
Greenbanks: State-level Environmental Review
Programs and their Associated Costs, Westview
Press, Boulder, Colorado
Healy, R.G. (1979) 'Environmental Regulations and
the Location of Industry in the U.S.: A Search
for Evidence', paper presented at a
Conference on The Role of Environmental and
Land Use Regulation in Industrial Siting, The
Conservation Foundation, Washington, DC
Kneese, A.V. and Schultze, C.L. (1975) Pollution,
Prices and Public Policy, Brookings Institu-
tion, Washington, DC
Köhn, R.E. (1974) 'Industrial Location and Air
Pollution Abatement', Journal of Regional
Science, 14, 55-63
Landsberg, H.H. (ed.) (1979) Energy: The Next
Twenty Years, Ballinger, Cambridge,
Massachusetts
Leone, R.A. (1976) Environmental Controls: A Study
of the Impact on Industries, Lexington Books,

Lexington, Massachusetts
Magee, M.L. and Cooper, H.B.H. (1978) An Evaluation
    of the Restrictiveness of Texas Air Control
    Board Regulation V on Control of Volatile
    Organic Compound Emissions from Petroleum
    Storage and Other Facilities, Environmental
    Study 1, Center for Energy Studies, University
    of Texas, Austin, Texas
McKee, H.C. (1978) 'The Problem of Equity in
    Emission Offsets', Journal of the Air Pollution
    Control Association, 28, 602-3
Organisation for Economic Co-operation and
    Development (1979) The State of the Environment
    in OECD Countries, OECD, Paris
O'Riordan, T. (1976) Environmentalism, Pion, London
Pearson, C. and Pryor, A. (1978) Environment:
    North and South, Wiley, New York
Pennwell Publishing Co. (1980) International
    Petroleum Encyclopaedia 1980, Pennwell, Tulsa,
    Oklahoma
Powelson, E.J. (1978) 'Industrial Wastewater Dis-
    charges on the Houston Ship Canal: A Study of
    Regulation and Water Quality Improvement',
    unpublished MA thesis, University of Texas,
    Austin, Texas
Radian Corporation (1977) The Impact of the Clean
    Air Act on Expansions of Petroleum Refining,
    Petroleum Storage and Selected Petrochemical
    Feedstock Facilities (2 volumes), Radian
    Corporation, Austin, Texas
Rosenbaum, N. (1976) Land Use and the Legislatures,
    Urban Institute, Washington, DC
Rothenberg, J. (1974) 'The Physical Environment' in
    J.W. McKie (ed.), Social Responsibility and
    the Business Predicament, Brookings Institu-
    tion, Washington, DC, pp. 191-216
Royston, M.G. (1979) Pollution Prevention Pays,
    Pergamon Press, Oxford
Russell, C.S. and Spofford, W.O. (1973) Residuals
    Management in Industry: A Case Study of
    Petroleum Refining, Johns Hopkins University
    Press, Baltimore
Shefer, D. and Guldman, J.M. (1975) 'A Model of Air
    Quality Impact on Industrial Land-use
    Allocation' in E.L. Cripps (ed.), Regional
    Science - New Concepts and Old Problems, Pion,
    London, pp. 66-83
Smith, T.T., Jr (1977) 'Prevention of Significant
    Deterioration - Discussion Paper', Journal of the
    Air Pollution Control Association, 27,
    849-52

Stafford, H.A. (1977) 'Environmental Regulations and
    the Location of U.S. Manufacturing: Specula-
    tions', Geoforum, 8, 243-8
Stafford, H.A. (1979) Principles of Industrial
    Facility Location, Conway Publishing, Atlanta,
    Georgia
Stern, A.C. (1977) 'Prevention of Significant
    Deterioration - A Critical Review', Journal of the
    Air Pollution Control Association, 27, 440-53
Storper, M., Walker, R. and Widess, E. (1981)
    'Performance Regulation and Industrial
    Location: A Case Study', Environment and
    Planning A, 13, 321-38
Tannahill, G.K. (1976) 'The Hydrocarbon/ozone
    Relationship in Texas', paper presented at a
    meeting of the Southwest Section, Air Pollution
    Control Association, Dallas, Texas
Texas Air Control Board (no date) A Summary of Air
    Pollutant Emissions in Texas, 1973, Emissions
    Inventory Section, Austin, Texas
Thompson, R.G., Calloway, J.A. and Nawalanic, L.A.
    (eds.) (1978) The Cost of Energy and a Clean
    Environment, Gulf Publishing Co., Houston
Tolley, G.S. (1975) 'The Resource Allocation Effects
    of Environmental Policies' in E.S. Mills (ed.),
    Economic Analysis of Environmental Problems,
    National Bureau of Economic Research, New York,
    pp. 133-63
US Congress (1969), National Environmental Policy
    Act of 1969. Public Law 91-190, 91st Congress,
    2nd Session, Section 101(a)
US Environmental Protection Agency (1974) Develop-
    ment Document for Effluent Limitations
    Guidelines and New Source Performance
    Standards for the Major Organic Products
    Segment of the Organic Chemicals Manufacturing
    Point Source Category. Washington, DC
US Environmental Protection Agency (1976) No Small
    Task: Establishing National Effluent Limita-
    tions Guidelines and Standards, Washington, DC
US Environmental Protection Agency (1979a)
    'Revisions to the National Ambient Air Quality
    Standards for Photo-chemical Oxidants'.
    Federal Register, 44, 8202-21
US Environmental Protection Agency (1979b)
    'Emission Offset Interpretative Ruling',
    Federal Register, 44, 3274-99
Walter, I. (1976) International Economics of
    Pollution, Macmillan, London
Watts, H.D. (1980) The Large Industrial Enterprise,
    Croom Helm, London

Wett, T. (1978) 'Ethylene Market Faces Overcapacity',
     <u>Oil and Gas Journal</u>, 76, No. 36, 63-8
Zurick, D.R. (ed.) (1971) <u>Water Wasteland</u>, Center
     for the Study of Responsive Law, Washington, DC

Chapter 6

THE ROLE OF THE STATE IN REGIONAL DEVELOPMENT

Thomas A. Clark

Only recently have geographers begun to formulate a
view of the state as playing a dual role in regional
systems, shaping and being shaped by the complex
processes of interdependent development (Dear and
Clark, 1978; Carney, Hudson and Lewis, 1980).
Already, however, it can be demonstrated that such a
perspective is indispensable for understanding
regional development in most advanced and many newly
emerging nations.   Conceptions of the state as a
mere adjunct of the national space-economy, as
immutable, as benign, or as an apolitical advocate
of the general welfare, are inadequate.   Rather it
must be seen at least potentially as a vehicle for
the expression of powerful interests (Fainstein and
Fainstein, 1978a), and as a central institution of
considerable reach (ACIR, 1980-1; Johnston, 1980),
capable of sustaining major national and multinational corporations (Holland, 1976; Pred, 1976;
1977; Malecki, 1979), of managing internal affairs,
and of governing relations with other nations,
relations which may produce differential effects
among internal regions (McConnell, 1980).
     Most state actions have regional implications
(Stohr and Todtling, 1977; Parr, 1979; Richardson,
1978b), but regional policy has proven difficult to
prescribe (Domazlicky, 1978; Gaile, 1979; Lande,
1977) or assess (Folmer, 1980; Pred, 1976; Clark,
1979).   Indeed actual benefits from regional policy
may be illusory at times (Choguill, 1977; Friedmann
and Weaver, 1979; Hinderink and Sterkenburg, 1978;
Rodwin, 1978; Molle, 1980).   Most major non-
regional policies, however, do have significant
spatial consequences, favouring some regions and not
others.   These include monetary (Mathur and Stein,
1980), taxing (Greytak and McHugh, 1978; Zimmerman,
1980), and spending policies (Gough, 1975;

Wohlenberg, 1976; Johnston, 1978; 1979a; 1980;
Government Research Corporation, 1976; Glickman,
1980), and still other policies affecting international migration, investment and trade.   Most of
these address particular classes of interests or
segments of society which are not uniformly distributed over space.   There is also evidence of analogous impact from the selective withdrawal of
national support systems under conservative
administrations as in both the United States and
Great Britain (Johnston, 1979b; Mackay, 1979;
Keeble, 1980).   A robust representation of the
state is therefore a critical ingredient in any
theory professing to explain regional development
or to guide regional policy.
     The ensuing commentary seeks a conceptual
synthesis portraying the role of the state in the
regional development process.   It is restricted to
advanced capitalist nations, though broader
relevance is likely.   Three categories of regional
theory now address the evolution of regional income
disparities in these nations: neoclassical,
cumulative causation and marxian constructs.
Following a brief characterisation of each, and an
evaluation of its capacity to explain the evolution
of regional welfare disparities in advanced nations,
its treatment of the state will be assessed.   From
this assessment will be derived a conceptual
synthesis.   The United States, though not wholly
representative, is used as the major empirical
example.   All three regional theories examined are
found to be inadequate in their treatment of the
state.   Each ignores or oversimplifies the
mechanisms through which fiscal, monetary, service-
providing, foreign relations and rule-making
functions of government affect regional systems.
They also fail to show how conflicts arising within
the interconnected hierarchies of contending
classes and places translate into state actions, and
how regional policies and national policies
mutually affect and transform one another.

Regional Process and the State

An adequate representation of the state in regional
development under advanced capitalism requires
three main ingredients.   First, a basis of motivation for state actions must be established.   Why
does the state take the actions it does?   How do
established state institutions and procedures affect

state actions and their results?   How may these
institutions themselves become objects of contention,
and how may changes in institutions and laws lead
to changes in the nature of state activity and in
the structure of the economy and society?

Second, the instruments of state action which
may influence geographical processes must be
identified.   The functions and related instruments
of the state are diverse, and as a result may be
independently motivated.   Adapting Dear and Clark's
summary (1978) of literature, four distinct
functions of the state can be identified: (1)
supplier of goods and services, (2) manager or
facilitator of the market-place, (3) manipulator of
basic social values and aspirations, and (4)
'arbiter' among competing social groups.   These
functions, however, are not wholly independent.
Certainly the first is common to most nations,
whereas the third may at times be incidental to the
second and fourth.   In addition, the state's role
as arbiter may take any of a variety of forms
depending on the configuration of competing social
interests, the range of policy responses it commands,
and the distribution of power between public and
private sectors and among regions and social classes.

Anticipating later discussion of pertinent
regional theories, it is clear that each represents
only some of these state functions.   In standard
neoclassical theory, the state's prime function is
as manager of the marketplace, while its role as
supplier of social capital is given considerable
attention in Mera's augmented neoclassical construct
(1975), and in both cumulative causation (Myrdal,
1957) and marxian unequal exchange theories
(Emmanuel, 1972; Liossatos, 1980).   For Mera,
however, state intervention in private market processes is an exogenous act rather than a product of
political process.   Early versions of cumulative
causation, as well as marxian theory are more
successful in supplying a rationale for state
behaviour.

The state's function as supplier of goods and
services is of considerable importance in shaping
patterns of regional development and entails both
taxing and spending impacts.   Through these
activities the state is joined to the major resource
circuits (i.e. flows of commodities and money) within the interregional economy.   An adequate representation of these activities should separate the
ways in which public revenues are generated in
order to assess their distributional consequences.

Likewise, it will be necessary to disentangle the
several distinct categories of spending including
(1) social transfers, (2) subsidies to corporations
in the form of write-offs, bail-outs and positive
incentives, (3) direct investments by government in
capital facilities, infrastructures and government
administration, and (4) procurements.   Taken
together these expenditures, balanced against
revenue flows, constitute a major influence on the
relative welfare of regions.   O'Connor (1973)
further observes that sustained fiscal imbalances
may themselves become instruments for structural
change when the state can no longer perform its
customary functions.   In the United States, reductions in federal subsidies to urban governments can
be expected in the near future to foster fiscal
crises in some regions which will ultimately induce
both population displacements and possibly also new
types of inter-jurisdictional, intra-regional fiscal
arrangements.
     The third requirement for an adequate portrayal
of the state's role in regional development is a
conceptual representation of how government policies
impact upon the interregional system leading to
subsequent socio-economic and ultimately political
responses.   These, of course, then become the basis
for successive rounds of state action since the
state is responsive to private sector power shifts
set in motion by an accumulation of state actions.
The state must be regarded as an endogenous agent,
an intrinsic element of social and spatial process.
These three requirements constitute the standard
by which existing regional theories will be judged,
and they will be incorporated in the final synthesis.
Before assessing the treatment of the state in
existing theories, however, their implications for
temporal variations in regional welfare disparities
will be considered.

Regional Welfare Disparities: Theoretical Perspectives

While empirical testing of the assumptions underlying each of the theories to be examined has proven
difficult, it is possible to deduce from each theory
expectations regarding the direction of change over
time in the per capita income gap between leading
and lagging regions.   These expectations can be
matched against actual trends in individual nations.
Neoclassical theory, founded on the seminal, well-

known work of Borts and Stein (1964), and recently
elaborated by Richardson (1973), Smith (1975) and
Mera (1975) among others, provides a rationale for
the convergence among subnational regions of wages
in individual industries, and of average wages when
industrial structure is constant among regions, in
the absence of agglomerative advantages.   Wage convergence, it is noted, is not equivalent to per
capita income convergence unless dependent populations are distributed proportionate to employment,
and no other types of personal income exist.

Cumulative causation theory, initiated by
Myrdal (1957), partially elaborated by Kaldor (1970),
and later reformulated in a more rigorous but
mechanistic manner by Dixon and Thirlwall (1975) and
still later by Richardson (1978a), yields a radically
different result.   Under cumulative causation,
regional per capita income differentials are
normally thought to grow larger over time in the
absence of state intervention, as agglomerative
advantages associated with large firms, industry
complexes and regional scale foster rates of
regional income growth which exceed those of population in leading regions.   Indeed, Dixon and
Thirlwall (1975) argue that under cumulative causation regional differentials in product-output growth
rates will be constant, and that the same will hold
true for growth rates of per capita earned incomes,
leading to divergence.

In contrast, marxian theory in general (Clark
and Dear, 1981) espouses an historical-materialist
conception of the state, in which its structure,
function and purpose are rooted in a socio-spatial
dialectic.   Further, some argue that regional disparities are intrinsic to capitalist development
(Mandel, 1976; 1978) and are necessary though not
sufficient for the survival of capitalism
(G.L. Clark, 1980).   Marxian treatment of regional
disparities usually incorporates a process of core-
periphery polarisation similar to that conceived
by cumulative causation theory.   Whereas the latter
stresses the ameliorative potential of state action
in diminishing spatial disparities generated by
market forces,however, marxists assign to the state
a central role in fostering such disparities.   Both,
however, admit that the state may also intervene to
lessen spatial disparities.   For marxists its
motivation, arising out of the social relations of
production, would be to foster social justice or
avert threats to the established political order
(also Hirschman, 1958).   It must be stressed,

Thomas A. Clark

nevertheless, that marxian analysis usually assigns
to spatial organisation a secondary, often deriva-
tive, role within the transformational dialectics
of class struggle (Soja, 1980).

Empirical Evidence of Regional Welfare Disparities

Empirical evidence of the evolution of regional welfare disparities within nations can at best give
only indirect support for the theories under review
since a match between theoretical expectations and
actual evidence may occur even when underlying
assumptions are invalid.   Available evidence, in
fact, raises considerable doubt regarding the
general relevance of any simple model of regional
process.   Customarily, temporal change in the level
of regional inequality has been measured by comparing the degree of per capita income dispersion
among regions at successive times within single
nations, and also by comparing this dispersion
across nations at different stages of development.
Williamson (1965), using both approaches, claimed
to have found considerable support for his assertion
that regional income inequalities would rise during
early stages of national development, then fall as
a result of economic and institutional forces active
during industrialisation, irrespective of all other
differences among nations at similar stages of
evolution.   Recently, however, this generalisation
has been confronted by contrary evidence reported
by Gilbert and Goodman (1976), and by the author
(in El-Shakhs et al., 1980, Table 10 and Figure 2).
It seems that regional inequality is neither a
simple function of national development, nor is it
wholly independent of such factors as the degree of
urban primacy, institutional structure and economic
base.
      Similarly, Kuznets' (1955) parallel assertion
that aggregate income inequality among individuals
would increase, then decrease, over the course of
national development has now also been called into
question by recent findings (Wright, 1978) indicating that institutional and other factors may be of
overriding importance in income distribution.   Per
capita regional income, in any event, is but one
measure of welfare (Richardson, 1978a, p.23), and
is not entirely representative (Smith, 1973).
Further, perceptions of inequality may have more
effect on regional aspirations (von Böventer, 1975)

174

and development than does the mere reality of
measurable differentials (Reiner, 1974).

The evolution of regional inequalities in the
United States, on which this empirical examination
will focus, conforms quite closely to the Williamson
hypothesis. In fact, longitudinal analysis of
this nation was one major basis for his conclusions.
Once regional inequalities in the United States have
been described, the pertinence of alternative
theories of regional development for this advanced
capitalist nation will be assessed.

Overall, Williamson (1965) observed that the
coefficient of variation (standard deviation divided
by the mean) of per capita personal income between
states increased steadily from 1840 to 1880,
followed by a significant deceleration of this
divergent trend from 1880 to 1920, and subsequently
a decline in inequality which was only momentarily
halted during the Great Depression and which
accelerated after the mid-1930s. Convergence over
the last half-century, Keuhn (1971) has noted, has
been due primarily to the fact that the constant
dollar growth rate of the mean value of state per
capita incomes exceeded that of their standard
deviation. He therefore concluded that while
relative convergence had indeed occurred, absolute
disparities have actually increased. Recently,
however, it has been demonstrated that absolute
disparities also began to decline over the last two
decades (Clark, 1982).

Williamson's data, however, conceal the
separate contributions of earned (participatory) and
other (non-participatory) income components to the
regional convergence of per capita personal incomes
in the United States over the last five decades.
Further disaggregation would reveal how each component has performed, and possibly suggest underlying distributive mechanisms. Wage and salary
income is now the largest participatory component,
and accounts for about two-thirds of total personal
income. Its performance is therefore crucial in
shaping the pattern of regional income disparities
both directly and by acting through regional
savings to generate property returns (rents,
dividends and interest payments).

Using an information statistic to assess
relative inequality, Semple (1977) reports overall
convergence since the 1930s of regional incomes per
worker for each major industry in the United States,
except agriculture (see also Gaile, 1977; Coelen,
1978). Disaggregation of this component by industry

175

and occupation suggests that not all subcategories
are equally prone to converge (see also Perin and
Semple, 1975).   Lande and Gordon (1977), for
example, have assessed temporal trends in wages per
production worker man-hour among states in seventeen
manufacturing industries (see also Lipietz, 1980).
In only about half of the industries examined did
wages grow more rapidly from 1947 to 1967 in low-
than in high-wage regions.   Original analysis by
the author using Survey of Current Business data
(US   Department of Commerce, 1956; 1962; 1973;
1980) indicates that the remaining participatory
components behaved irregularly during the 1930-79
period (Clark, 1982).   Nevertheless, the coefficient
of variation in state per capita participatory
income as a whole has fallen steadily, indicating
convergence (Table 6.1).

Non-participatory income components include
both returns to property and government transfers
(including Federal old-age, survivors, disability
and hospital insurance; workmen's compensation;
veterans' assistance; food stamp payments;
supplementary security income and related programmes
- less personal contributions to these transfer
programmes).   These are also of major interest
because the former relates to the capital-owning
segment of society, and the latter reveals the
impact of government social transfer programmes.
In 1930, the regional distribution of each was more
unequal than was the distribution of participatory
income.   Thus the addition of property returns and
gross transfers to the participatory income base
actually increases the degree of aggregate inequality in this year (Table 6.1).   From 1930 to
1979, however, the coefficients of variation of
both state per capita property and gross transfer
incomes declined steadily after small increases in
the 1930s, indicating convergent trends.   By 1979,
the level of regional inequality of both property
and gross transfer income had fallen substantially.
In this year, gross transfer income had a coefficient of variation equal to that of per capita
participatory income, while property's was much
higher.   Property returns, though, account for only
one-eighth of total pre-tax personal income so their
impact on the cumulative coefficient of variation
is proportionately small, as is that of gross
transfers (also one-eighth of total pre-tax
personal income).

Further, individual payments for social
insurance amount to just one-third of gross social

Table 6.1: Coefficients of Variation of US State
Per Capita Personal Income, Cumulative by Source,
1930 to 1979

		Type of income[a]		
Year	Participatory	Plus Property Income	Plus Gross Transfers	Less Personal Contributions to Social Transfers
1930	.332	.386	.382	.381
1940	.314	.362	.359	.355
1950	.236	.254	.243	.247
1960	.200	.215	.207	.208
1970	.154	.158	.149	.148
1979	.129	.131	.117	.119

Note: a. Successive, cumulative additions to participatory
income.   Participatory and property incomes are pre-tax.
Source: Author's calculations based on data from United
States Department of Commerce, 1956, 1962, 1973 and 1980.

transfers so their impact on the cumulative coefficient of variation (Table 6.1) is also slight.
Still, by 1979, the coefficient of variation of net
per capita transfers was well above that for gross
transfers, indicating the spatial distribution of
per capita contributions to social insurance
programmes is now less equal than is the spatial distribution of per capita benefits.   Government's
impact on the distribution of property and participatory incomes among regions, furthermore, is
probably even more substantial.   This impact,
however, is far more difficult to document since the
bulk of government expenditures affects incomes indirectly.
      Cross-sectional analysis of the correlations
between each per capita non-participatory component
and the remaining income components gives additional
insight into the inter-connectedness of income flows
in the United States.   Examining the temporal trend
in key correlations may suggest underlying process
(Table 6.2).   Overall, there has occurred a substantial weakening of the statistical association
between per capita property and gross transfer

Thomas A. Clark

Table 6.2: Cross-sectional Correlations of US State
Per Capita Non-participatory Income Components,
1930, 1950, and 1979

Attribute[b]	Property Income			Gross Transfers		
	1930	1950	1979	1930	1950	1979
Population	.43	.21	.19	.30	.33	.38
State Per Capita Income:						
Total	.93	.86	.44	.79	.43	.28
Wage & Salary	.90	.79	.44	.76	.48	.17
Other Labour	.64	.65	.20	.61	.29	.17
Proprietors'						
Farm	-.49	-.24	.27	-.12	-.31	-.41
Non-farm	.72	.55	.45	.83	.47	-.21
Property				.67	.33	.19
Gross Transfers	.67	.33	.19			
Social Payments[c]	.20	.62	.23	.08	.56	.10

Note: a. Pearson's r, N=48, such that r is significant at
.10 (two-tail test) if $|r| \geq 24$.

    b. For same year as column head.

    c. Personal contributions to social transfer programmes.

Source: Same as Table 5.1.

incomes across states since 1930. In addition, with
few exceptions each of these classes of income has
become steadily less associated with the various per
capita participatory components over the last fifty
years (Table 6.2). While these data do not permit
definitive conclusions, they offer strong circum-
stantial evidence that participatory, property and
transfer incomes are independently distributed, and
that explanatory models of regional process must
treat each of these dimensions more explicitly than
is now customary.
    So aggregate a characterisation of regional

process, of course, masks the underlying dynamics of regional income convergence in the United States over the last half century. This period witnesses many confounding trends which in retrospect are difficult to disentangle. Numerous urban centres rose to prominence in outlying regions (Pred, 1977), subnational and international spatial systems underwent accelerated integration, major shifts in population occurred (Berry and Silverman, 1980; Vining and Kontuly, 1978), the national space-economy increasingly emphasised the service-performing industries (T. Clark, 1980; Beyers, 1979; and Lipietz, 1980) and regional industrial structures became more homogeneous (Sternlieb and Hughes, 1976). Even these more disaggregate trends underlying regional convergence, however, reflect and sustain more fundamental transformation of society and its key institutional frameworks. Most importantly, the period in question was marked by the massive expansion of the federal establishment (ACIR, 1980-1), the emergent duality of competitive and monopoly capital (O'Connor, 1973), and a redefinition of the roles of public and private sectors (Holland, 1976; Harrison and Kanter, 1978).

Within this period of convergence, the state at times led, and at others lagged the private sector. The ascendance of monopoly capital interests in the late 19th and early 20th century culminated in the Great Depression, which stimulated, then co-opted, the demand-side palliatives of the New Deal. Subsequent expansion of the federal apparatus, however, was dictated by the balances of interests arising in an expanding number of policy arenas, and by the independent momentum of governmental institutions themselves (Markusen, 1979). Throughout the period of regional income convergence, non-regional sectoral politics have predominated, serving as the combat zone of special interests, but often the political outcomes have had spatial consequences (US Department of Commerce, 1970). Gradually, though, a regional consciousness has begun to emerge particularly with the surprisingly slow recognition that the growth of the South and West has occurred at the expense of the northern tier of states. To date, in any case, federal regional policy (Martin, 1979; Clark, 1978) has commanded few resources and therefore has had little impact, and even now there appears to be little support of regional policy despite perfunctory protests (President's Commission, 1980). The US experience is not wholly representative of all

advanced nations. Those having more pronounced
regional disparities, proportionately more
intrusive public sectors, lower per capita GNPs, and
greater spatial variation in class structure will
probably be more inclined to engage in spatial
politics and fashion regional policies (see also
Fainstein and Fainstein, 1978b).

In sum, none of the theories examined wholly
anticipates the character of regional convergence
in the United States. Of the three, neoclassical
theory is the one most often associated with
regional convergence. This theory, however,
deduces only the convergence of wages, and, as noted,
empirical findings of wage convergence are contingent on the degree of industrial aggregation.
Further, wage convergence among regions in
individual industries does not imply overall convergence in average wage rates when aggregated over
all industries because industrial structures may
differ among regions. Neoclassical theory also
anticipates the regional convergence of returns to
capital, but there is little evidence supporting
this proposition, at least in the United States,
due to the lack of capital stock data. Finally,
the equalisation of factor returns among regions -
premised by this theory - does not dictate the
regional patterns of non-participatory income components. And, in any case, per capita returns to
property (i.e. capital) in the preceding analysis
refer to the place of residence of persons receiving
this type of income, not the places where the income
is generated. In addition, the theory is mute
regarding the effect of direct government transfers
on personal incomes.

The two other theories examined would seem to
be somewhat more difficult to reconcile to conditions in the United States. Neither can be
rejected out of hand, however. Cumulative causation theory, in fact, may be reconciled, so long
as the process it describes is separated from the
initial patterns of spatial polarisation it normally
assumes. Most cumulative causation theorists have
studied nations whose economy is spatially polarised,
with one or a few urban centres dominating a vast
underdeveloped rural periphery. In contrast, the
United States has numerous major centres scattered
across it. Despite absolute size differences
among these leading centres, the larger ones may
actually be quite competitive, each positively
influencing personal incomes in its hinterland. In
the United States, therefore, the forces of

cumulative causation may promote simultaneous gains
in many rival urban centres, many of which subsequently induce complementary gains in immediately
surrounding areas through favourable 'spread'
effects including population dispersion (Vining and
Kontuly, 1978; see also Richardson, 1980). Mera
provides considerable implicit support for this contention, noting that in the United States (1975,
pp. 65-6):

> where the mobility of capital and labour is
> high, the growth rate of regions is largely
> determined by the rate of technical progress
> and the growth rate of social capital. The
> rate of technical progress is high in regions
> where the level of technology is low and the
> rate of urbanisation is high, but the growth
> rate of social capital is high where the income
> per capita is high. Therefore, these two
> factors are counterbalancing. Less-developed
> regions are growing primarily because of the
> growth of the technical level, while more
> developed regions are growing because of the
> growth of social capital. The net effect is
> determined by the rate of technical progress
> rather than by the growth of social capital.
> ... As far as the growth of technical progress
> is concerned, urbanisation explains it far
> better [emphasis added]. High growth rates
> of technical progress in the South Atlantic,
> West South Central and Mountain regions
> correspond well to large increases in the per-
> centage of the urban population in these
> regions. In the United States ... urbanisation
> is a significant determinant for improving
> production efficiency.

In this manner agglomeration can explain the process
of regional income convergence, serving both as a
vehicle for growth in lagging regions and as a basis
for further productivity gains beyond those achieved
through capital intensification and technical
progress during rapid urbanisation.

For Mera (1975), at least, the process is
straightforward. As urbanisation slows, investment
in social capital sustains growth by improving the
productivity of large supplies of initially underutilised urban labour. The effectiveness of this
process, however, may be limited due to (1) increased
competition from rival centres, (2) fluctuations in
labour demand which both attract and sustain

dependent populations of underutilised labour, and
(3) increasing political rigidity and polarisation
in metropolitan areas (due to suburbanisation,
political fragmentation, resource depletion and the
aging of infrastructure and capital).

Unfortunately, cumulative causation theory, and
Mera's supporting analysis even more so, constitute
disjointed explanations of process which tend to
confuse material manifestations of process such as
rate of urbanisation and level of agglomeration
with process itself. It is for this reason that
elements of the marxian perspective, itself not yet
well developed and hampered by excessive rigidity
in translating world and national processes into
regional constructs, may be useful. Marxian
analysis, however, is even more equivocal regarding
the evolution of regional disparities and their significance within the course of national development.
It does not indicate, specifically, whether disparities are essential for capitalist accumulation
and the reproduction of the social relations of production, or merely probable consequences of other
forms of unevenness in capitalist development.
While Markusen (1979) argues that 'nothing in the
logic of capitalist accumulation ... requires
spatial differentiation,' G.L. Clark (1980) asserts
that disparities may attend capitalist development
though the state may intervene in the interest of
capital at times either to promote or diminish
disparities when they are a barrier to accumulation.
Unequal exchange theory however does rather un-
equivocally indicate divergence, yet if the
potential for agglomeration exists in most regions
(as it does in the United States) then the previous
argument rationalising convergence under cumulative
causation may also be used to reconcile this theory
with the empirical reality of regional convergence
in the United States.

None of these theories however has an adequate
and complete conception of the state. Out of
these building blocks, nevertheless, elements may be
derived for a more realistic synthetic model. This
is the purpose of the final section of this paper.

The State in Regional Theory

Versions of each of the major classes of theory
under review address regional process in advanced
capitalist nations such as the United States.
While each can be reconciled to the empirical

reality of per capita regional income convergence in
this nation, as previously demonstrated, it remains
to be seen whether their underlying assumptions are
plausible. Neoclassical, then cumulative causation
and finally marxian theories will be examined from
this perspective. While each builds upon a wide
range of economic and institutional assumptions,
this review will stress treatment of the state since
it is posited in this commentary that political
processes associated with state action are crucially
important.

Neoclassical Theory: Standard and Augmented

The state in standard neoclassical theory (Borts,
1960) is submerged in, yet exogenous to, the free
market, though a more active role is implicit in
recent augmented versions of the model. Standard
theory asserts that factors of production will flow
among regions in response to price/wage differentials reflecting variance in marginal productivities
(mp). Key assumptions are the perfect inter-
regional mobility of capital and labour, finite
factor stocks, full factor employment, uniform output prices due to uniform production functions (with
no economies of scale) and zero transportation costs,
and a single production sector in all regions. It
then follows that the simultaneous movement of
factors will lead to factor price equalisation
(Figure 6.1, Part A), regardless of differences of
resource endowment which may have led initially to
disparity. Commodity circuits are implicit though
it is also assumed there will be zero cost in con-
verting output to capital goods.
     Suppose there are two regions, one having low
and the other high per capita income and wage rates
(assuming uniform labour participation rates).
The low-income region, assume further, has the
lower capital/labour (K/L) ratio and therefore a
higher marginal capital yield (mpK) and lower labour
productivity (mpL). If factors move from areas of
low to areas of high marginal yields, then labour
will migrate from the depressed region to the more
prosperous region. Simultaneously, capital will
flow in the reverse direction seeking higher yields
in the depressed region. The lower-income region
will consequently have higher rates of growth in the
capital/labour ratio, labour productivity and per
capita income (earned). Eventually, without
technical progress, factor returns among regions will

Figure 6.1:   The Neoclassical Model: Basic and Augmented

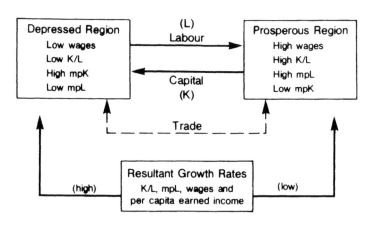

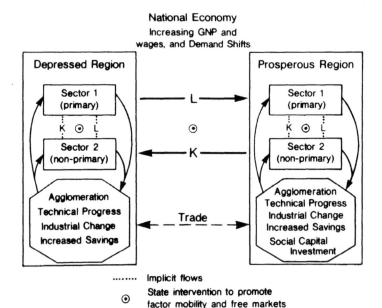

converge even when output scale and population size
differentials remain.   Factor price equalisation -
regional convergence - is therefore seen to arise
directly from simplistic assumptions.   Given full
employment of factors which are paid according to
their marginal products, the wage paid labour will
increase more rapidly in the depressed than in the
prosperous region.   This is the essential condition
for relative participatory income convergence.
    Early empirical tests of the standard model in
the United States (Borts, 1960; Borts and Stein,
1964), as previously noted, focused on wages in the
absence of capital stock data, and generally fail
to support it.   Of three intervals examined, Borts
found wage convergence at the state level in only
one (1929-48), a period in which demand conditions
promoted price increases in labour-intensive sectors
of growing states leading to increases in marginal
labour productivity - a condition unlikely to continue.   Borts and Stein (1964) later attributed
their failure to find evidence supporting the model
to undue sectoral aggregation, and demand differentials between high and low wage industries.   They
argued that population growth in high wage regions
induces expansion of their service sectors
supported by capital stock increases, and that at
the same time national growth rates in demand for
the output of high wage industries exceeds that for
low wage industries.
    Recently, others have sought to disaggregate
the standard neoclassical model.   Smith (1975)
departed from the standard assumption of identical
production functions in all regions by incorporating
technical progress.   His formulation is a
'variable proportion' growth model employing most
of the usual assumptions while disaggregating components of change in regional capital and labour.
Net regional investment is the sum of local savings
plus net capital inflow less depreciation, and
absolute change in a region's labour supply is the
product of the exogenous labour participation rate
and the sum of natural increase plus net migration.
Lande and Gordon (1977) relax the model assumptions
by admitting variable returns to scale.   In
addition, following the lead of Borts (1960) who
suggested industrial disaggregation (service and
non-service) in order to capture the effect of
demand differentials among industries, they disaggregate by industry - while retaining a supply-
side emphasis.   They empirically demonstrate convergence conditions, finding that in the United

States marginal products of capital fell and labour rose with increases in the capital/labour ratio in sixteen of seventeen industries examined (p.63). It is concluded that wage growth rate differentials among regions can be explained by both industry-mix and agglomerative factors, and that neoclassical assumptions do not inevitably lead to aggregate wage convergence.

Mera, by contrast, departs even more substantially from the standard neoclassical model (1975, pp. 40-5). He assumes the existence of not two but three production factors: (1) capital, which is perfectly mobile and responsive to productivity differentials, (2) labour, which is also perfectly mobile in response to 'utility' not wage differentials, and (3) social capital, which is allocated exogenously according to government policy. He argues, however, that wage rates may not equalise among regions due to: (1) price differences in untradeable goods, (2) differentials in unmarketable residential amenities, (3) quality and productivity differentials in the labour force partly due to (4) disparities in levels of social capital (inducing productivity differentials in labour) and technology (inducing productivity differentials in capital). Thus, while Borts (1960) stressed the demand side of the service sector, Mera stresses the supply side of social capital investment - education and other human services - and thereby ascribes to government a new role in neoclassical theory beyond the maintenance of free markets.

An augmented version of the neoclassical model would then include several production sectors, agglomeration effects, technical progress, a more disaggregated treatment of the saving-investment function in capital accumulation, and a full representation of demand-side dynamics including commodity circuits and demand shifts (Figure 6.1, Part B). In addition, factor flows between sectors of single regions would also be required though their treatment in the literature is cursory (Lande and Gordon, 1977, pp. 67-8). Further, neoclassical formulations always regard state intervention as exogenous. That is, nothing indicates what conditions within the system would motivate state action to maintain free markets or to generate social capital. Ultimately this issue must be addressed.

The role of the state would have to be substantial if the free market assumptions of neoclassical theory are to be supported. The tendency toward the concentration of wealth and power in the form

of industrial conglomerates within and among sectors
will not be limited without the active involvement
of the state, yet the state may in fact become subservient to the interests of major industrial concentrations.   Clearly, however, disaggregation of
production sectors, recognition of labour quality
differentials, and acknowledgement of the state
functions of market regulation and service provision
are first steps in modelling the inter-regional
economic system required if the state is to be conceived as an endogenous participant in the development process.

The State in Cumulative Causation

The theory of cumulative causation occupies a middle
ground between neoclassical and marxian formulations, sharing key features with each.   In Myrdal's
well-known original version (1957), the state played
a highly visible role, whereas in later more
rigorous statements of the theory, this role has
been misplaced entirely or subsumed within vaguely
articulated conceptions of agglomerative advantage
(Kaldor, 1970; Dixon and Thirlwall, 1975; and
Richardson, 1978).

For Myrdal, whatever the cause of initial
regional disparities, both capital and labour will
tend to flow between regions in response to price/
wage differentials arising from differences in
agglomerative scale among regions, seeking the
highest possible returns.   With trade liberalisa-
tion and the industrialisation of core regions these
movements will reinforce the agglomerative advantages of more prosperous regions.   Core-periphery
polarisation during this period may actually be
fostered or sustained by actions of the state - that
is 'all organised interferences with the market'
(1957, p.42) - which accumulate capital extracted
from peripheral regions for investment in core
regions while stifling dissent on the part of the
poor and peripheral.

As national development progresses, the scope
of potential state action broadens and at least in
democratic societies, the objectives of state
policies may change.   At a very low level of
economic development private market forces will predominate in the spatial polarisation of the nation,
Myrdal argues (p.41), and these inequalities will
actually inhibit economic development and weaken
the 'power base for egalitarian policies'.   At a

higher level of development societal forces will
(p.41)

> strengthen the spread effects (see also Gaile,
> 1980) and tend to hamper the drift towards
> regional inequalities; this will sustain
> economic development, and at the same time
> create more favourable conditions for policies
> directed at decreasing regional inequalities
> still further. The more effectively a
> national state becomes a welfare state -
> motivated in a way which approaches a more
> perfect democracy, and having at its disposal
> national resources big enough to carry out
> large-scale egalitarian policies with bearable
> sacrifices on the part of regions and groups
> which are relatively better off - the stronger
> will be both the urge and the capacity to
> counteract the blind market forces which tend
> to result in regional inequalities; and this,
> again, will spur economic development in the
> country ... in circular causation.

Myrdal does not make it completely clear though why
the democratic state would be motivated to promote
regional and class equality, and in particular does
not explore thoroughly the relative contributions of
political liberalism and self-interest on the part
of the owners of capital in the shift from the
'oppressor' state to the 'welfare' state. Indeed,
class-based co-optation and coalition are at best
implicit, and the realignment in space of social
class by regional change is scarcely addressed. As
a consequence no clear indication is provided of the
complex interdependence of class conflict and
regional competition.

A plausible, rudimentary explanation of the
emergence of redistributional policy, however, can
be pieced together from fragments provided by
Myrdal (1957), Hirschman (1958), Friedmann (1972)
and Siebert (1969). This follows:

1. Initially core-based capital may seek
   investment outlets in peripheral regions
   even as labour is drawn to core area
   employment opportunities. In addition
   incipient regional trade reveals peripheral
   market potentials.
2. To protect their investments in peripheral
   areas and insure market expansion for their
   core-produced products, core-based capital

      owners promote limited state action to up-
      lift the periphery.

3.   These efforts selectively to involve the
      poor and peripheral in the core-based
      economy lead to gradual liberalisation of
      the rules of democratic participation, a
      process spearheaded by the more easily
      mobilised urban working class. At first
      only secondary benefits accrue to the
      periphery.

4.   The previously disenfranchised thereby
      secure a larger voice in national parlia-
      ments and at first can only realise gain
      through the recognition of interests shared
      with core-based owners of capital.
      Fledgling capital interests located in the
      periphery may ally with this grouping so
      long as class polarisation remains sub-
      merged within the developmental opportuni-
      ties of the periphery.

5.   Through this process the state advances
      from its position as an adjunct of capital
      to a role as nexus of political contention
      among unequal competitors and to a later
      role as redistributional agent. Its re-
      distributive role - building on its
      precursor in capturing wealth for capital
      investment - evolves as powers once distri-
      buted among lower levels of government
      accrue to the political centre.

Interregional deconcentration (spread), of
course, may at times arise as much from market
forces as state action. Hirschman (1958), for
example, argued that spread ('trickle-down') will
gradually increase so long as the periphery economy
complements the core's, allowing for the sale of
peripheral exports in core regions - though resulting surpluses may still be extracted by the core
economy. He also argued that core area congestion
would stimulate polarisation reversal with the onset
of scale diseconomies and congestion in core regions,
but here again the core area may well capture
resulting capital surpluses. Mera, however, asserts
that there is no convincing empirical evidence that
economic returns to public or private investments in
the United States are greater in medium and smaller-
sized cities than in larger ones (1975, pp. 17-34).
Urban dispersal nevertheless appears to be under way
currently in numerous advanced nations (Vining and
Kontuly, 1978) including the United States.

Thomas A. Clark

Figure 6.2:   The Role of the State in Cumulative
Causation

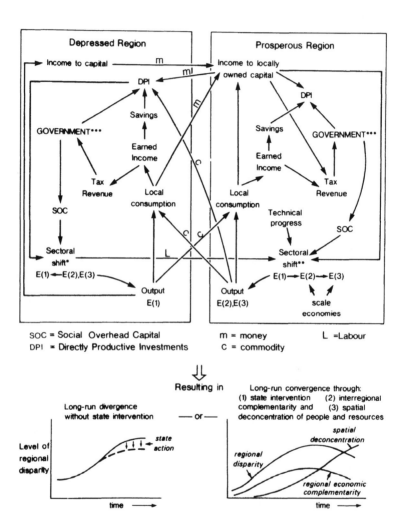

Notes: * Internal shift of capital and labour from
secondary (E(2)), and tertiary sectors (E(3)) to
primary sector (E(1)).
      ** Shift of production factors to E(2) and
E(3) from E(1).
      *** All levels of government.   Interregional
revenue and expenditures are implicit.

   Kaldor, among others, has sought to sharpen
the concept of cumulative causation (1970).   He
asserts that leading regions will be able to sustain
their advantage because of the Verdoorn effect.
The Verdoorn hypothesis is that leading regions will
have a cumulative advantage because manufacturing
tends to concentrate there, and manufacturing is the
sector most able to secure increasing returns to
scale (both internal and external).   Kaldor goes
still further, asserting that because competition is
perfect in agriculture and imperfect in the major
industries of leading regions, commodity and factor
flows will both favour leading regions.   Growth of
both investment and consumption within leading
regions, furthermore, will largely depend on growth
in external demand for their exports (see also
Berentsen, 1978).
   Dixon and Thirlwall (1975), in turn, have
sought to elaborate Kaldor's thesis.   They conclude
that the model would seldom produce divergence in
output growth rates among regions as most cumulative
growth theorists claim, but rather constant growth
rate differentials sustained by the Verdoorn effect.
The relationship between output growth rates and
regional _per capita_ incomes is unfortunately
ambiguous.   Workers are paid a money wage in this
model equal to the efficiency wage (average cost of
labour per unit of output) multiplied by the
average product of labour.   Increases in either
the efficiency wage or the product of labour will
increase the money wage.   Assuming labour participation rates are uniform among regions, the rates
of growth in regional _per capita_ income and in money
wages (per worker) will,be proportionate.   Consequently the growth rate of _per capita_ income and
regional output will also be positively related.
Thus both relative and absolute interregional _per
capita_ income divergence will occur so long as the
regions having higher output growth rates also have
higher initial incomes.   Constant output growth
rate differences would yield constant differences
in _per capita_ income growth rates, and therefore
income divergence, but three caveats are in order:
(1) interregional migration may dampen this effect
(Ledent and Gordon, 1980), (2) non-participatory
personal income (transfers and capital returns) is
assumed zero in all regions, and (3) income divergence may be the result of key assumptions built
into the model rather than intrinsic to the process
itself.   Clearly, the state plays no visible role
in this formulation, but the model's strengthening

Thomas A. Clark

of the structural and parametric basis of cumulative
causation advances it well beyond Myrdal's more
institutional formulation of the process. The
Dixon-Thirlwall (1975) formulation, however, is only
a partial portrayal of the circuit of flows in
cumulative causation.

In order to portray the system of resource flows
which are explicit or inherent in these statements
of cumulative causation, a flow-chart has been drawn
(Figure 6.2). In this chart, money (m), commodities (c) and labour (L) flow between a depressed
and a prosperous region. Savings and capital returns are invested in directly productive investments
(DPI) while the state (national and regional)
extracts money through taxation to generate social
overhead capital (SOC). Technical progress, scale
economies, a superior capacity to accumulate
capital and sectoral shifts foster high output
levels in the prosperous region (PR). Growth is
hampered in the depressed region (DR) due to
capital extraction by the PR, loss of skilled
labour to the PR, diminished competitive capacity of
secondary and tertiary industries, money drain to
imports, and low savings potential because of low
incomes. Returns to capital will accrue in the PR,
inflating non-participatory income there. In
addition, wage rates will also grow more rapidly
in the PR so long as the wages in the secondary and
tertiary industries of the PR increase. This
requires that the influx of workers from (1) the
PR's primary sector, (2) the DR, and (3) the less
competitive secondary and tertiary industries within
the PR, will not dampen growth rates in aggregate
wages. The resulting disparity graphs shown at the
bottom of Figure 6.2 show Myrdal's and Hirschman's
expectations, respectively.

The State in Unequal Exchange: Marxian Formulations

The state has always been central in marxian analysis (see, for example, Sweezy, 1968), though the
spatial organisation of development within nations
has generally been treated as more consequential
than causal (Soja, 1980). There are, however,
important exceptions. Some have adopted a core-
periphery paradigm, treating depressed regions as
repositories of production factors and market potential which may be selectively drawn into the sphere
of core-based capitalist production and commodity

192

circulation at critical moments by state actions on behalf of the owners of capital. Some, indeed, argue that regional disparities ensure the survival of capitalism (Mandel, 1976), and that capitalism both determines and is in turn shaped by spatial distributions of people and resources (Lefebvre, 1976). Soja supports this contention, seeking fashion a socio-spatial dialectic which recognises the tension between class and region (Soja, 1980, p. 219):

> ... there exists in the structure of organised space under capitalism a fundamental relationship which is homologous to and inseparable from the structure of social class, with each structure forming part of the general relations of production. This connection between the vertical and horizontal dimensions of the class struggle in both theory and practice requires a fundamental rethinking of many Marxist concepts to introduce a more explicit spatial problematic ...

A key departure point for several marxian analysts is the distinction Mandel makes between differentiation and equalisation in regional profit rates under capitalism (1978, p. 43). During early, competitive industrial capitalism, he argues, profit rates will tend to equalise among regions and sectors - an assertion virtually the same as that of neoclassical theory. During late capitalism, which is marked by the emergence of giant corporate conglomerates,however, expanded production will depend upon the extraction of 'superprofits', and these in turn will foster sectoral and regional disparities (see also Harvey, 1975). The bifurcation of capital, between competitive and monopolist sectors, in advanced capitalism might therefore be expected to promote counter-thrusts within the overall process of spatial development.

Emmanuel (1972) has proposed one of the more complete marxian models of regional process under capitalism stressing the regional and structural consequences of exogenous interregional wage differentials favouring core regions. Though originally formulated to examine international relations when capital but not labour is mobile among nations, a variant of the model may be relevant at the subnational level. Permitting labour mobility in this setting might be expected to have little impact on regional outcomes. This model, summarised in Figure 6.3, Part A, takes wages as exogenously given and higher in the core region. The gross return to capital is total profit less the cost of labour, and though product prices may be somewhat inflated by monopolists in order to pre- serve an acceptable profit margin, labour is the major determinant of prices.

Through interregional trade, surplus in the form of capital returns will accrue to the prosper- ous region because (1) monopolist industries in the core will secure external capital through the sale of exports, and (2) capital drain from the core due to the purchase of the depressed region's exports will be less than commensurate. In addition, the core region, home of numerous monopolist industries, will also gain a cumulative advantage over the depressed region due to (1) internal market expan- sion (population increase due partly to inmigration, plus higher per capita buying power), (2) extraction by the state of a portion of wealth in circulation for investment in social capital to promote greater labour productivity, (3) scale economies made possible by growth in output and an expansion in internal and external markets, and (4) the influx of foreign capital (Emmanuel, 1972, pp. 130-7). The process of core area development is cumulative and ensures increasing disparity between core and depressed regions.

Unequal exchange, as it has been presented, offers little insight into the systemic nature of the state, either as an agent promoting change or as it responds structurally within the marxist dialec- tic. Yet this is the essence of the historical materialist conception of the capitalist state (Holloway and Picciotto, 1978). As shown in Figure 6.3, Part B, the state is positioned struc- turally between capital (competitive and monopolist) and labour (differentiated in numerous dimensions of place, skill, aspiration and wage). In the marxist conception the state functions predominantly as an agent manipulating labour in the interest of capital, though it also may be thought of as serving a mediating role between capital and labour when labour can command the instruments of power. Capital and labour, furthermore, join in the social relations of production. Labour is exchanged for products (this yields wages paid labour and surplus value extracted by capital). Labour unions, generally more prominent in urban core regions with more highly trained workforces, and other organisa- tions also mediate between labour and capital in the

Figure 6.3:   The State in Unequal Exchange Theory

A. Unequal Exchange

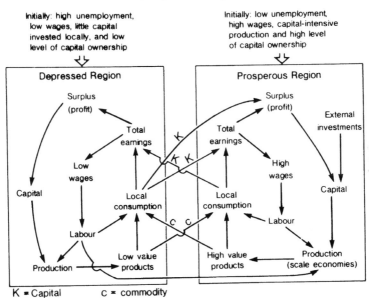

Initially: high unemployment, low wages, little capital invested locally, and low level of capital ownership

Initially: low unemployment, high wages, capital-intensive production and high level of capital ownership

K = Capital   c = commodity

B. The Labour -Capital Relationship in Capitalist Nations

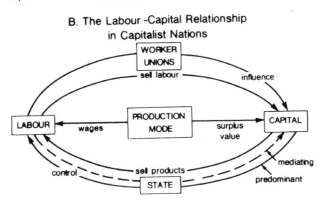

Thomas A. Clark

Figure 6.4:   The Political Economy of Regional
Development: A Synthesis

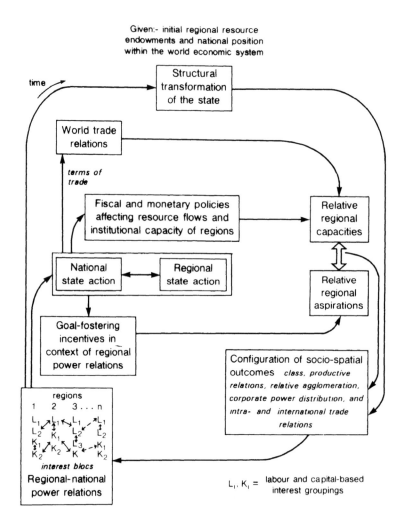

interests of labour.   Over time the state seeks to
legitimise the existing social relations of production while at the same time easing accumulation
crises by mobilising labour and capital reserves as
uneven development progresses.   It will further
seek to maintain - reproduce - the existing social
and perhaps spatial relations of production in an
effort to sustain the capitalist framework.

A number of further points should be made to
conclude this review.   First, the cumulative causation model is similar in several respects to both
the neoclassical model and the marxian model of unequal exchange even though the latter two differ
substantially.   Second, Myrdal's cumulative causation model is a loosely organised verbal construct
claiming almost universal relevance irrespective of
the nature of the political economy.   Neoclassical
models assume free market economies while in
marxian analysis the state and the social relations
of production are artifacts of a process sustained
by a dialectical progression arising out of contradictions inherent in the regimen of change itself.
For marxists, the opposition of classes is the
prime cause of change, but in the theory of cumulative causation the relative position of the classes
is governed by less clear-cut processes of economic
transformation.

Synthesis: Regional Development and the State
The preceding evaluation indicates important
strengths and weaknesses in existing regional
theories and suggests the necessity of selective
synthesis.   This section provides such a synthesis,
characterising regional process in relation to two
embedded loops of circular causation.   The inner
loop corresponds to short-run processes of regional
change, whereas the outer loop corresponds to longer-
run structural transformations stimulated in part by
spatial and sectoral tensions generated within the
stable institutional framework of the inner loop
(Figure 6.4).   In this conception, power relations
run both vertically within sectoral interest groupings, and horizontally across space.   Neither labour
nor capital is monolithic.   Indeed mutual gain is
possible through intra-regional coalition arising
from regional allegiances and the internalisation of
intra-regional externalities.   Further, these
groupings may establish distinctive methods of
political involvement at each level - local, regional,

Thomas A. Clark

national - of political participation.   The state,
however, is not the only medium of political
exchange.   In short, this conception departs from
the more simplistic, highly aggregated representation of class orientations in marxist theory and the
early expressions of cumulative causation.   It
recognises tensions within both labour-based and
capital-based elements of the marxian dichotomy, and
admits the duality of competitive and monopolist
segments of capital.   Further, while the state may,
in general, favour capital-based interests, tensions
within capital may well promote contradictions among
distinct facets of state action (see also Holland,
1976, chapter 6, and G.L. Clark, 1981).
     Regional-national power relationships establish
the political context to which the state responds.
They do so, not by promoting an independent,
'rational' course of action, but by shaping state
responses which reflect the various balances of
power associated with the substantive facets of
policy.   As shown in Figure 6.4, these governmental
responses impact upon regions through two interacting paths of causation.   On one hand, national
fiscal and monetary initiatives affect regional
capacity, while on the other, national subsidies to
regions may influence regional aspirations (see
also von Böventer, 1979).   'Capacity' is determined
not only by the net stock of resources available
to the region, but also by their distribution among
internal power groupings.   Capacity is multi-
dimensional, and varies among regions, and among
economic sectors within regions.   Regional aspirations, of course, may be subordinated to the
increasing multiplicity of internal interests,
though how regional aspirations form and affect
action is not well understood.   A further element
of state action bearing on regional capacity is
international trade policy, which has a strong
influence upon region-based multinationals, capacity
to attract foreign capital, and international competition for domestic markets.   The interaction
between regional capacities and regional aspirations
determines the interregional distribution of development (Figure 6.4).   This becomes the basis for
successive rounds of circular and potentially
cumulative interactions.   It is within this cycle,
repeatedly traversed, that the socio-spatial
dialectic unfolds.   The outer loop encompasses
other cumulative tensions which may lead to important
structural transformations in governmental institutions, changes which can subsequently alter the

198

process of political competition. In this way
institutions may attempt to keep pace with the
process of social transformations which their
previous actions and policies have helped to
engender.

   While previous discussion has established the
broad institutional basis for a synthetic model of
regional development within an interregional system
of power relationships and governing institutions
(Figure 6.4), the interpenetration of state fiscal
flows and interregional resource circuits remains
to be considered. This pattern of flows (Figure
6.5) is critical in articulating not only primary
points of tax and spending impact,but also the
secondary consequences which could be expected to
ensue. It should be immediately clear that models
such as Mera's (1975) and Emmanuel's (1972) which
exclude tax flows and all spending except for social
capital cannot possibly represent differential
regional growth when the state's fiscal involvement
is extensive in relation to the scale of the economy.

   In Figure 6.5, two regions are represented, one
depressed, the other prosperous. In each are
circuits of resource flows in the form of money and
commodity exchanges. In each, production generates
products (given regional industrial mixes which may
or may not be complementary) which are purchased
internally or as exports. Consumption is directly
affected by state action due to the price-effects of
state taxation as well as through state procurements
(P) and income transfers to unemployed and dependent
populations (T). Resultant revenues are paid to
labour in the form of wages (which the state may
further subsidise (T)),and to the state in the form
of taxes (X), with the remaining surplus going to
the owners of capital (which the state may further
subsidise (T)). Savings, capital surpluses and
direct government investments (I) are subsequently
spent on directly productive investments (DPI),
while the state may also invest in social overhead
capital (SOC) to appease labour to legitimate its
other actions (which may be perceived to benefit
capital owners), and thereby to foster productivity
increases.

   Throughout these resource circuits, there occur
key exchanges of money and commodities among regions,
and between corporations and the state. Fiscal
resources also flow among levels of the government
hierarchy. Exactly how resources will flow within
the national resource circuit will, of course,
depend on public and private sector decisions taken

Thomas A. Clark

in the context of the overall political economy of
development.   Several conclusions regarding the
role of the state in the development of interregional
systems arise.   First, there are many distinct
avenues of potential state involvement and these are
not always mutually reinforcing.   Each avenue of
involvement is affected by an accumulation of prior
policy decisions and by whatever sets of political
forces are currently operative.   Second, it may be
exceedingly difficult to specify the ultimate impact
of state fiscal policies.   This is because considerable 'leakage' among regions can occur through
interregional trade, capital extraction, and
external investment.   Third, there is a convincing
case to be made for a representation of state involvement within a circular and cumulative causation
framework.

Summary

The central issue addressed has been the role of the
state in regional development in advanced capitalist
nations.   Three ingredients for such a representation have been proposed.   These are a plausible
rationale for state action, a delineation of the
major instruments of state involvement in inter-
regional economic systems, and a description of how
state actions affect these systems and produce subsequent socioeconomic responses which lead in turn
to further state actions.   Three major classes of
regional theory were then assessed, including neoclassical, cumulative causation and marxian constructs.   Each was first evaluated with respect to
the compatibility of its predictions with the actual
evolution of regional income disparities in the
United States.   Versions of all three could be reconciled to the pattern of regional per capita
income convergence occurring in twentieth-century
America.   Next, the underlying assumptions of each
were examined, stressing the portrayal of the state
in regional process.   Each fell short when evaluated
against the three evaluative norms mentioned above.
Finally, a synthesis was offered, in which the state
is viewed as an endogenous agent in the development
of interregional systems.   This synthesis relates
the institution of the hierarchical state to the
dynamics of short-run intervention and long-run
structural transformation, also considering the
pattern of fiscal flows within interregional systems.
        Two conclusions arise from this inquiry.

Figure 6.5: The State in Regional Development under
Capitalism: Interregional Resource Circuits

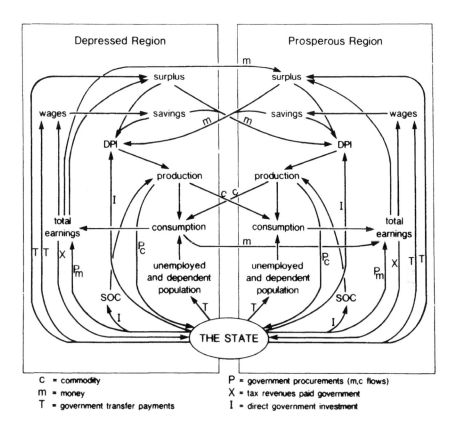

C = commodity	P = government procurements (m,c flows)
m = money	X = tax revenues paid government
T = government transfer payments	I = direct government investment

Thomas A. Clark

First, intra-national regional development is an
exceedingly complex process which is not likely to
be captured by any simple model. Second, the
successful representation of regional process in
any nation with a dominant public sector, as exists
in most advanced and many emergent nations, requires
a robust characterisation of the motivation,
function and spatial impact of the state. Articulating the state's role in regional development
should head the regional research agenda. It is
hoped that the preceding commentary will be of use
in shaping this important effort.

References

Advisory Commission on Intergovernmental Relations
    (1980-1) The Federal Role in the Federal
    System: The Dynamics of Growth (11 volumes),
    ACIR, Washington,DC
Berentsen, W.H. (1978) 'Regional Policy and
    Industrial Over-specialization in Lagging
    Regions', Growth and Change, 9, 9-13
Berry, B.J.L. and Silverman, L.P. (1980) Population
    Redistribution and Public Policy, National
    Academy of Sciences, Washington,DC
Beyers, W.B. (1979) 'Contemporary Trends in the
    Regional Economic Development of the US',
    Professional Geographer, 30, 34-44
Borts, G.H. (1960) 'The Equalization of Returns and
    Regional Economic Growth', American Economic
    Review, 50, 319-47
Borts, G.H. and Stein, J.L. (1964) Economic Growth
    in a Free Market, Columbia University Press,
    New York
von Böventer, E. (1975) 'Regional Growth Theory',
    Urban Studies, 12, 1-29
von Böventer, E. (1979) 'Development in Space:
    Historical Processes, Aspirations, Potential-
    ities and Inequalities', Papers, Regional
    Science Association, 42, 29-38
Carney, J., Hudson, R., and Lewis, J. (eds.) (1980)
    Regions in Crisis: New Perspectives in
    European Regional Theory, Croom Helm, London
Choguill, C.L. (1977) 'Regional Planning in the US
    and the UK: A Comparative Analysis',
    Regional Studies, 11, 135-45
Clark, G.L. (1979) 'Predicting the Regional Impact
    of a Full Employment Policy in Canada: A Box-

Jenkins Approach', Economic Geography, 55, 213-26

Clark, G.L. (1980) 'Capitalism and Regional Inequality', Annals of the Association of American Geographers, 70, 226-37

Clark, G.L. (1981) 'The Employment Relation and Spatial Division of Labor: A Hypothesis', Annals of the Association of American Geographers, 71, 412-24

Clark, G.L. and Dear, M.J. (1981) 'The State in Capitalism and the Capitalist State', in M.J. Dear and A.J. Scott (eds.), Urbanization and Urban Planning in Capitalist Societies, Methuen, New York, pp. 45-61

Clark, T.A. (1978) 'Regional Development: Strategy from Theory' in G. Sternlieb and J.W. Hughes (eds.), Revitalizing the Northeast, Center for Urban Policy Research, Rutgers University, New Brunswick, New Jersey, pp. 405-43

Clark, T.A. (1980) 'Regional and Structural Shifts in the American Economy Since 1960' in S.D. Brunn and J.O. Wheeler (eds.), The American Metropolitan System: Present and Future, Halsted, New York, pp. 111-25

Clark, T.A. (1982) 'Components of Regional Income Convergence', The Review of Regional Studies, (forthcoming)

Coelen, S.P. (1978) 'Regional Income Convergence/Divergence Again', Journal of Regional Science, 18, 447-57

Dear M.J. and Clark, G.L. (1978) 'The State and Geographic Process', Environment and Planning A, 10, 173-83

Dixon, R. and Thirlwall, A.P. (1975) 'A Model of Regional Growth Rate Differences on Kaldorian Lines', Oxford Economic Papers, 27, 201-14

Domazlicky, B.R. (1978) 'The Regional Allocation of Investment: A Neoclassical Model', Regional Science Perspectives, 8, 2, 1-13

El-Shakhs, S., et al. (1980) 'National and Regional Issues and Policies in Facing the Challenges of the Urban Future' in Population of the Urban Future, United Nations Fund for Population Activities, New York

Emmanuel, A. (1972) Unequal Exchange: A Study of the Imperialism of Trade, Monthly Review, New York

Fainstein, N.I. and Fainstein, S.S. (1978a) 'Federal Policy and Spatial Inequality' in G. Sternlieb and J.W. Hughes (eds.), Revitalizing the Northeast, Center for Urban Policy Research, Rutgers University, New Brunswick,

Thomas A. Clark

New Jersey, pp. 205-28

Fainstein, S.S. and Fainstein, N.I. (1978b)
    'National Policy and Urban Development',
    Social Problems, 26, 125-46

Folmer, H. (1980) 'Measurement of the Effects of
    Regional Policy Instruments', Environment and
    Planning A, 12, 1191-202

Friedmann, J. (1972) 'A General Theory of Polarized
    Development', in N.M. Hansen (ed.), Growth
    Centers in Regional Economic Development, Free
    Press, New York, pp. 82-107

Friedmann, J. and Weaver, C. (1979) Territory and
    Function: The Evolution of Regional Planning,
    University of California Press, Berkeley

Gaile, G.L. (1977) 'Effiquity: A Comparison of a
    Measure of Efficiency with an Entropic Measure
    of the Equality of Discrete Spatial Distribu-
    tions', Economic Geography, 53, 265-82

Gaile, G.L. (1979) 'Alternative Spatial Investment
    Strategies', Economic Geography, 55, 227-41

Gaile, G.L. (1980) 'The Spread-Backwash Concept',
    Regional Studies, 14, 15-25

Gilbert, A.G. and Goodman, D.E. (1976) 'Regional
    Economic Disparities and Economic Development:
    A Critique' in A.G. Gilbert (ed.), Development
    Planning and Spatial Structure, Wiley, New
    York, pp. 113-41

Glickman, N.J. (1980) The Urban Impacts of Federal
    Policies, Johns Hopkins Press, Baltimore

Gough, I. (1975) 'State Expenditure in Advanced
    Capitalism', New Left Review, 92, 53-92

Government Research Corporation (1976) 'Federal
    Spending: The North's Loss is the Sunbelt's
    Gain', National Journal, 26, 878-91

Greytak, D. and McHugh, R. (1978) 'The Effects of
    Federal Income Taxation and Inflation on
    Regional Income Inequalities', Journal of
    Regional Science, 18, 57-70

Harrison, B. and Kanter, S. (1978) 'The Political
    Economy of States' Job-creation Business
    Incentives', Journal of the American Institute
    of Planners, 44, 424-35

Harvey, D. (1975) 'The Geography of Capitalist
    Accumulation', Antipode, 7, 9-21

Hinderink, J. and Sterkenburg, J.J. (1978) 'Spatial
    Inequality in Underdeveloped Countries and the
    Role of Government Policy', Tijdschrift voor
    Economische en Sociale Geografie, 69, 5-16

Hirschman, A.O. (1958) The Strategy of Economic
    Development, Yale University Press, New Haven

Holland, S. (1976) Capital Versus the Regions,

Macmillan, London

Holloway, J. and Picciotto, S. (eds.) (1978) State and Capital, Edward Arnold, London

Johnston, R.J. (1978) 'Political Spending in the United States: Analyses of Political Influences on the Allocation of Federal Money to Local Environments', Environment and Planning A, 10, 691-704

Johnston, R.J. (1979a) 'Congressional Committees and the Inter-state Distribution of Military Spending', Geoforum, 10, 151-62

Johnston, R.J. (1979b) 'The Spatial Impact of Fiscal Change in Britain: Regional Policy in Reverse', Environment and Planning A, 11, 1439-44

Johnston, R.J. (1980) The Geography of Federal Spending in the United States of America, Wiley, New York

Kaldor, N. (1970) 'The Case for Regional Policies', Scottish Journal of Political Economy, 17, 337-48

Keeble, D.E. (1980) 'Industrial Decline, Regional Policy and the Urban-Rural Manufacturing Shift in the United Kingdom', Environment and Planning A, 12, 945-62

Keuhn, J.A. (1971) 'Income Convergence - a Delusion', Review of Regional Studies, 2, 41-51

Kuznets, S. (1955) 'Economic Growth and Income Inequality', American Economic Review, 45, 1-28

Lande, P.S. (1977) 'A Re-examination of Regional Growth Models and Development Policies', Regional Science Perspectives, 7, 67-9

Lande, P.S. and Gordon, P. (1977) 'Regional Growth in the United States: A Re-examination of the Neoclassical Model', Journal of Regional Science, 17, 61-9

Ledent, J. and Gordon, P. (1980) 'A Neoclassical Model of Interregional Growth Rate Differences', Geographical Analysis, 12, 55-67

Lefebvre, H. (1976) The Survival of Capitalism, St. Martin's Press, New York

Liossatos, P. (1980) 'Unequal Exchange and Regional Disparities', Papers, Regional Science Association, 45, 87-104

Lipietz, A. (1980) 'Inter-regional Polarization and the Tertiarisation of Society', Papers, Regional Science Association, 44, 3-18

Mackay, R. (1979) 'The Death of Regional Policy - or Resurrection Squared?', Regional Studies, 13, 281-95

Malecki, E.J. (1979) 'Locational Trends in R & D by

Large U.S. Corporations, 1965-1977', Economic
    Geography, 55, 309-23
Mandel, E. (1976) 'Capitalism and Regional Dis-
    parities', Southwest Economy and Society, 1
Mandel, E. (1978) Late Capitalism, Verso, London
Markusen, A. (1979) 'Regionalism and the Capitalist
    State: The Case of the United States',
    Kapitalistate, 7, 39-62
Martin, R.C. (1979) 'Federal Regional Development
    Programs and U.S. Problem Areas', Journal of
    Regional Science, 19, 157-70
Mathur, V.K. and Stein, S. (1980) 'Regional Impact
    of Monetary and Fiscal Policy: An Investigation
    into the Reduced Form Approach', Journal of
    Regional Science, 20, 343-51
McConnell, J.E. (1980) 'Foreign Direct Investment
    in the United States', Annals of the
    Association of American Geographers, 70, 259-70
Mera, K. (1975) Income Distribution and Regional
    Development, University of Tokyo Press, Tokyo
Molle, W. (1980) Regional Disparity and Economic
    Development in the European Community, Saxon
    House, Farnborough, Hampshire
Myrdal, G. (1957) Economic Theory and Under-
    developed Regions, Duckworth, London
O'Connor, J. (1973) The Fiscal Crisis of the State,
    St. Martin's Press, New York
Parr, J.B. (1979) 'Spatial Structure as a Factor in
    Economic Adjustment and Regional Policy', in
    D. MacLennan and J.B. Parr (eds.), Regional
    Policy, Martin Robertson, London, pp. 191-210
Perin, E. and Semple, R.K. (1975) 'Recent Trends in
    Regional Income Inequalities in the United
    States', Regional Science Perspectives, 6, 65-
    85
Pred, A.R. (1976) 'The Interurban Transmission of
    Growth in Advanced Economies: Empirical
    Findings Versus Regional Planning Assumptions',
    Regional Studies, 10, 151-71
Pred, A.R. (1977) City-systems in Advanced Economies,
    Wiley, New York
President's Commission for a National Agenda for the
    Eighties (1980) Urban America in the Eighties,
    Washington, DC
Reiner, T.A. (1974) 'Welfare Differences within a
    Nation', Papers, Regional Science Association,
    32, 65-82
Richardson, H.W. (1973) Regional Growth Theory,
    Macmillan, London
Richardson, H.W. (1978a) 'The State of Regional
    Economics', International Regional Science

Review, 3, 1-48

Richardson, H.W. (1978b) 'Growth Centers, Rural Development and National Urban Policy: A Defense', International Regional Science Review, 3, 133-52

Richardson, H.W. (1980) 'Polarization Reversal in Developing Countries', Papers, Regional Science Association, 45, 67-86

Rodwin, L. (1978) 'Regional Planning in Less Developed Countries: A View of the Literature and Experience', International Regional Science Review, 3, 113-31

Semple, R.K. (1977) 'Regional Development Theory and Sectoral Income Inequalities' in J. Odland and R.N. Taaffe (eds.), Geographical Horizons, Kendall/Hunt, Dubuque, Iowa, pp. 45-63

Siebert, H. (1969) Regional Economic Growth, International Textbook, Scranton, Pennsylvania

Smith, D.M. (1973) The Geography of Social Well-being in the United States, McGraw-Hill, New York

Smith, D.M. (1975) 'Neoclassical Growth Models and Regional Growth in the United States', Journal of Regional Science, 15, 165-81

Soja, E. (1980) 'The Socio-spatial Dialectic', Annals of the Association of American Geographers, 70, 207-25

Sternlieb, G. and Hughes, J.W. (eds.) (1976) Post-Industrial America: Metropolitan Decline and Inter-regional Job Shifts, Center for Urban Policy Research, Rutgers University, New Brunswick, New Jersey

Stohr, W. and Todtling, F. (1977) 'Spatial Equity - Some Antitheses to Current Regional Development Doctrine', Papers, Regional Science Association, 38, 33-53

Sweezy, P. (1968) The Theory of Capitalist Development, Monthly Review, New York

US Department of Commerce, Bureau of Economic Analysis (1973) Survey of Current Business, 53, Tables 6-61

US Department of Commerce, Bureau of Economic Analysis (1980) Survey of Current Business, 60, Table 3

US Department of Commerce, Economic Development Administration (1970) 'Federal Activities Affecting Location of Economic Development', II

US Department of Commerce, Office of Business Economics (1956) 'Personal Income of States since 1929', Supplement to the Survey of Current Business

Thomas A. Clark

US Department of Commerce, Office of Business
    Economics (1962) Survey of Current Business,
    42, Tables 6-61
Vining, D.R., Jr, and Kontuly, T. (1978) 'Population
    Dispersal from Major Metropolitan Regions',
    International Regional Science Review, 3, 49-73
Williamson, J.G. (1965) 'Regional Inequalities and
    the Process of National Development', Economic
    Development and Cultural Change, 13, 3-45
Wohlenberg, E.H. (1976) 'Interstate Variations in
    AFDC Programs', Economic Geography, 52, 254-66
Wright, C.L. (1978) 'Income Inequality and Economic
    Growth', Journal of Developing Areas, 13, 49-66
Zimmerman, H. (1980) 'The Regional Impact of Inter-
    regional Fiscal Flows', Papers, Regional
    Science Association, 44, 137-48

Chapter 7

INSTITUTIONAL EFFECTS ON INTERNAL MIGRATION

Robin Flowerdew

Research on internal migration tends to have a
strong identification with particular disciplines:
economists have explained it in terms of responses
to wage and unemployment rates, geographers in terms
of distance-decay models, sociologists in terms of
community ties and life-style preferences.   All of
these approaches have merit, but neglect the fact
that for many people migration is regulated by the
institutional structures that affect their lives.
Most household heads who migrate between regions
end up with higher wage or salary levels, not
because they chose to move to a region with high
average wages, but because their move was a promotion to another job in a different branch of their
organisation, or because they applied successfully
for a widely-advertised job with another organisation.   Migration in particular professions may
reflect the tendency of those professions to reward
experience in different environments, and the degree
of acceptance in one place of qualifications gained
elsewhere.   Another important factor may be the
ease of obtaining suitable housing in a different
region, which may be affected by the organisation of
the real estate industry.   In Britain, most local
authority housing is allocated from a waiting list,
so in-migrants are likely to have severe difficulty
in finding housing of this type.   Government may
have policy directly relevant to migration, such
as encouragement to workers seeking jobs in a new
region, or planned expansion and overspill schemes.
Its policy may indirectly affect migration, by
encouraging or discouraging large employers from
locating in particular places.   Local differences
in welfare programmes may also have substantial
effects.
          This paper argues for the importance of studying

the effects of actions by governments, job-providing organisations, and other institutions on inter-regional migration. Such an approach may both illuminate the background to individual migration decisions, and improve our understanding of aggregate relationships of migration to distance and other variables.

Evidence from interview surveys (e.g. Johnson, Salt and Wood, 1974) indicates that the circumstances of an individual or household migration decision are essential to an understanding of that decision, and that frequently these circumstances are created or influenced by institutional actions. This is perhaps most directly the case where migration is accompanied by job change within organisations. American evidence (Long and Hansen, 1979) indicates that 23.8 per cent of all interstate migrant household heads move as a result of job transfers. For male household heads between 35 and 54, 34.0 per cent move for this reason. Many of the second largest category of household heads migrating between states, the 23.6 per cent taking new jobs or looking for work (unfortunately not distinguished in the data), can also be regarded as moving for reasons connected with institutional actions, while 5.4 per cent and 4.8 per cent respectively moved in connection with education or the armed forces.

Where a migrant is moving to take up a new job, institutions concerned with the worker's job search and the employer's search for labour will have played an important role in setting the preconditions for the job. Other relevant factors may include the availability of housing: institutions controlling the housing market, such as mortgage-granting institutions, may affect the possibility, or the destination, of a move. In Britain, for example, Johnson, Salt and Wood found that many people were unable to get a mortgage for house purchase in London, but were able to do so in other parts of the country.

Attempts to fit a regression model to inter-urban migration flows can usually achieve a reasonable fit on the basis of size and distance variables. Residuals from a model of this type (see Flowerdew and Salt, 1979) were found to be dominated by flows which could reasonably be accounted for in terms of institutional effects. These included the impact of movement between military bases, flows reflecting the recruiting policies of large organisations, and flows consequent on the expansion of a large company to a new location. It can be argued (Gleave and Palmer, 1980) that the route to improved modelling and pre- diction of inter-regional migration lies in disaggre- gation of migration flows by occupation, and an analysis of these disaggregated flows in relation to the operation of the labour market in these occupa- tions. These labour markets are greatly influenced by their institutional structures.

Research on Institutional Influences and Migration

There have been several strands of research con- ducted in recent years concerned with the relation- ship of migration and institutional action. In this section, some of these topics are outlined, including the effects of government policy, commercial and religious organisations, and educa- tional institutions on migration. The most intensively studied aspect of government action concerned with migration has been the impact of economic policy. There appears to be a general effect that inter-regional migration rates are higher in times of national prosperity than in a depression (Clark, 1982). Regional economic policy, however, may have more specific effects on the size and composition of specific migration streams.

Neoclassical economic theory stresses the role of migration in equalising per capita incomes between regions. A region of relative prosperity should have a labour deficit, resulting in wages being higher than in a depressed region, where labour surplus forces wages down. This condition should be rectified because the divergence of wage rates increases the incentives for inter-regional migration, resulting in labour migration from regions of surplus to regions of deficit. This theory is supported by an examination of regional net migration rates, but not by an examination of gross regional flows (Gleave and Cordey-Hayes, 1977). High-income regions, as predicted, have higher in- migration rates than low-income regions, but, in contrast to the theory, they also have higher out- migration rates. In addition, regional inequali- ties clearly persist despite the availability of migration as a potential equalising factor.

If governments wish to equalise prosperity between regions, therefore, some additional policy measures are necessary. Most of these policies are concerned with providing incentives for employers to locate in depressed regions and disincentives for

them to locate in prosperous regions. It is of
some importance to determine whether such policies
have an impact on migration. In particular, if new
industry is located in an area, what proportion of
the workers are recruited locally, and how many
migrate from elsewhere? Are there differences
between unskilled and skilled workers, or between
blue-collar and white-collar workers, in their migration rates and in their responsiveness to these
policies? Similarly, in times of recession,
governments may need to make decisions concerning
whether or not to discourage plant closures, and
the impact on local unemployment may be an important
consideration. The nature of this impact will be
modified if workers made redundant by plant closure
migrate to regions with better employment prospects.
Government policy on plant closures and, more
generally, public expenditure cuts, should be considered in a study of migration, just like the
impact of industrial location and regional development policy.

Other attributes of regions may be important
in a migration decision, and some of these may be
directly influenced by the policies of national and
local governments. Cebula (1979) discusses the
effects of differences between states in the USA in
their average welfare benefits, finding support for
the idea that higher levels of benefit deter in-
migration of whites and attract blacks. He also
reviews a number of other papers on the relationship between welfare payments and migration.
Cebula (1974) has also studied the impact of state
and local government tax policies on migration.
Greenwood and Sweetland (1972) used per capita
government expenditure as a predictor variable in
a regression model of inter-SMSA migration, and
other studies (Cebula, 1974; Ostrosky, 1978) have
concluded that higher levels of spending on education are attractive to migrants. Barsby and Cox
(1975) investigated the impact of public sector
variables on migration of the elderly, generally
concluding that levels of state old age benefits and
health care have only slight impact. Gober (1981)
suggests that Federal retirement benefits have made
retired people more financially secure, and hence
more able to migrate to areas which are attractive
for retirement. Projected changes in the US
Social Security system might well reduce this
effect.

Another type of government policy designed to
tackle regional inequalities is concerned directly

with migration, through instituting schemes intended
to help people migrate to areas of better economic
opportunity. Most European nations have schemes
of this type (reviewed by Klaassen & Drewe, 1973);
in Sweden, for example, one in five of gainfully
employed migrants receivesa removal grant (Öberg
and Oscarsson, 1979). The recent report on Urban
America in the Eighties has argued for a major
initiative of this type in the USA (US President's
Commission for a National Agenda for the Eighties,
1980). Studies have examined the operation of some
of these schemes, such as the British Resettlement
Transfer Scheme (Beaumont, 1976b; Johnson and Salt,
1980a). Interesting aspects of this work include
the necessity of publicising a scheme widely before
it can be effective, and the seeming abuse of the
programme by migrants who do not appear to have sought
employment following their assisted move, or who
returned to their area of origin after a short
government-assisted vacation (Beaumont, 1976a).
Problems may also arise, as Johnson and Salt (1981)
suggest in the British context, from conflicts
between these policies and other government actions,
such as housing legislation.

In addition to government actions, migration
may be encouraged by public or commercial organisations. In Britain, the National Coal Board (NCB)
has a scheme for assisting employees to move from
an area where mining is being phased out to an area
where it is expanding. The British Steel Corporation operates a similar scheme (Johnson and Salt,
1980a). Taylor (1969) has examined the factors
influencing participation in the NCB scheme.
Other studies have examined the impact of the relocation of industry (Mann, 1973) and of offices
(Bateman, Burtenshaw and Hall, 1971) on the migration of their employees. McKay and Whitelaw (1977)
were able to infer the importance of inter-
organisational transfers in Australia from the
migration rates of those occupational groups
associated with large organisations. Evidence on
the importance of personnel transfers in explaining
return migration in Japan has been provided by
Wiltshire (1979). Johnson and Salt (1980b) use
survey data to identify the age and occupational
characteristics of a subsample of migrants who moved
within the same organisation, finding that a high
proportion of these people received substantial
assistance from the organisation. Much survey and
anecdotal information is also available on the
mobility experiences of executives in large

Robin Flowerdew

organisations (Birch and Macmillan, 1971).

Historically, migration has sometimes been
associated with religious or ethnic institutions.
The most spectacular example in the USA may be the
movement of Mormons to Utah.   As Meinig (1965,
p. 198) reports, 'The Church leadership selected
the time and the place and many of the specific
persons for important new colonizations ... The
Mormon term for such selection is "called", and it was
tantamount to an order, though one usually willingly
obeyed as a proper duty whatever the personal
sacrifice.'   Other studies of the development of
ethnic communities have addressed similar issues.

Both commercial organisations and governments
have been active in sponsoring large-scale population movements.   Settlement on the American frontier
in the nineteenth century was promoted by government
action, especially the Homestead Act, and was often
sponsored by organisations - sometimes by railroads,
for which in-migration to an area was essential for
their capital investment to pay off.   In the USSR,
government has also been responsible for migration:
the Organized Labor Recruiting Service (Organobor)
has resettled 28 million workers mostly on a short-
term basis according to Lewis and Rowland (1979).
They also discuss Soviet agricultural resettlement
schemes, stating, however, that organised migration
has become much less important since 1950.

An interesting special case of institutional
effects on migration is that of student migration.
Desbarats (1977) has discussed the effects of the
centralised UCCA system on movements of British
students to universities, noting the relative rare-
ness of flows between England and Scotland as a
result of the differences in their educational
systems.   Comparison with American experience
(Scipione, 1973) shows the importance of differences
in national policies (British student grants including allowances for students to live away from home);
differences in tuition levels for in-state and out-
of-state students at many American institutions also
have obvious effects on migration patterns.   Such
institutional effects may be important not only in
themselves, but may have influences on the subsequent residence of the students concerned.

Although work on migration done to date from
an institutional perspective has explored only a few
isolated themes there are signs that more systematic
attention is being paid to these factors.   Although
his Resource Paper on 'Changing Migration Patterns
within the United States' is primarily behavioural

in approach, Roseman stresses the importance of studying institutions to understand new migration patterns (1977, p. 31):

> migration streams may be a function of institutional decisions. In a society with a continuing large governmental employment sector, the location decisions for major federal employment installations, military and otherwise, will have important impacts upon overall migration patterns. The decisions may be at the whims of pork-barrel legislators or other government decision-making structures which have had little concern as yet for impacts upon population distribution.

In the next section, attention is directed to some of the most important research areas where an institutional approach might be particularly fruitful.

Institutional Factors and their Effects

Organisational Policies and Career Patterns

One of the most important factors in accounting for particular moves is the intra-organisational transfer (Johnson and Salt, 1980b, provide an excellent discussion of the issues reviewed in this section). In Johnson et al.'s survey (1974), 28 per cent of a sample of labour migrants (people who had changed both home and job) had the same employer before and after moving. In a period where large multi-locational organisations are becoming increasingly important in capitalist economies, moves within an organisation may become important for three reasons. First, the organisation may need to adjust its operations, which may be done by moving personnel from contracting to expanding plants or, more specifically, by moving personnel in certain technical or managerial categories to plants where their skills were more in demand. Secondly, it may have a policy of transferring employees between plants, through internal promotion or through a self-conscious management development policy designed to give some employees experience of many aspects of the organisation's operations as a training for top management positions. A third factor is the organisation's willingness to transfer employees to other locations at their own

request.   In Johnson et al.'s survey (1974), 31 per
cent of migrants who did not change their employer
had asked for the move, mostly because they wanted
to live in the new area.   This situation could
easily arise where one spouse had obtained a job in
a new area and the other wishes to find employment
in the same place.
        When an organisation expands, contracts,
rationalises, or moves its operations, employee
transfer is likely to be selective in favour of
managerial and skilled workers.   This is because
these workers are the ones in whom the organisation
has the biggest investment; they may also be the
hardest to hire in the new location.   Migrants may
differ in their ability to move with their employers;
Mann (1973) found that three-quarters of male
employees moved with Bird's to their new plant in
Banbury, but only one-quarter of female employees
moved.
        Survey information on people who moved at the
request of their employers is available from
Johnson and Salt (1980b).   Offers of promotion
accounted for 41 per cent of these moves, with a
further 22 per cent being offered a job elsewhere -
other reasons included the shift of the whole office,
experience or training, and the conclusion of a job.
The employees moved by their employers tended predominantly to be managerial, executive and professional workers, and to include a higher proportion
of people from the older age-groups than is the
case for migration generally.
        Three-quarters of the movers in the survey
reported by Johnson and Salt (1980b) received
assistance from the employers, with 94 per cent of
managers and executives who had not initiated the
move receiving aid.   As Johnson and Salt point out
(p. 283), 'The migration policies of organisations
thus seem to be geared to helping those who already
can move easily surmount the barriers to labour
migration.'   Removal expenses and legal expenses
were the most common types of assistance given,
although many intra-organisational migrants received
a disturbance allowance, and expenses for travel or
temporary accommodation.
        As Johnson and Salt (1980b) indicate, several
factors may affect migration within an organisation.
These include the internal structure of the organisation.   Obviously, a company which operates only
in one or a few regions of a country will only move
employees within this relatively limited territory.
An organisation whose internal structure is broken

down into regional divisions is also less likely
to sponsor inter-regional movement. If the internal structure of the organisation is based on
product divisions, however,internal promotions or
transfers may be far more likely to necessitate long-
distance migrations. Organisational structure and
geographical distribution of operations may therefore be important variables affecting the amount and
nature of intra-organisational migration.

A second type of consideration is the policy
adopted by large organisations towards career
development, whether such a policy has been
formally agreed upon or has just developed in response to specific organisational needs. Does the
organisation have a training programme for management trainees that involves geographical mobility?
How often does it move its employees around? Are
all managerial level employees moved, or just an
elite, or others in non-managerial grades? Is
there any common sequence of moves, perhaps alterna-
ting moves between headquarters and regional offices,
or a series of moves from the smallest centres
gradually ascending both the urban and the corporate
hierarchy? Such policies may differ significantly
between companies and organisations of different
type, and they may be subject to change through
time.

One particularly important factor may be the
reaction of employees to the experience of being
moved around, with the concomitant disruption to
home life, friendships, spouse's job and children's
schooling. Anecdotal evidence suggests that
refusals to move are increasingly prevalent
especially because of the growing number of two-
career households (Rapoport and Rapoport, 1971).
Organisational transfer policy may have to adjust to
individual preferences for stability.

Professional Careers

Professional workers, like executives, have very
high mobility rates, although there is a great deal
of variation between professions (McKay and
Whitelaw, 1977, p. 38) Haber (1981) suggests
several reasons for this, including the low cost of
interarea job search relative to lifetime earnings,
greater resources for financing such search, a taste
for mobility acquired through university education,
and access to sources of information about non-local
jobs through professional societies and journals.

The idea of 'career pattern' as an influential factor in determining migration propensities was developed by Leslie and Richardson (1961) and Ladinsky (1967b).   According to this idea, migration is most likely to take place in association with career changes.   These career changes, such as entry into the labour force and promotions, are likely to be concentrated at particular ages.   In some careers, promotion or a better job may involve geographical mobility - dependent on the structural characteristics of the job concerned.

Within the professions, an important factor influencing migration rates may be the extent to which the professional has fixed capital in a location.   If he or she is dependent on community recognition and trust, a good reputation and a regular clientele may take some time to build up, and a move would require starting again from nothing.   If professional practice necessitates heavy expenditure on special buildings or equipment, migration is again unlikely.   These types of professionals are often self-employed; salaried professionals, working for industry or government, on the other hand, may be highly mobile (Ladinsky, 1967a).   Even in professions of the former type, however, migration may not be unusual, in the early stages of a career, where young workers may assist in well-established businesses before starting out on their own.

In some professions, there are characteristic patterns of migration, with a high degree of mobility expected in the early stages of a career, as in the British medical professions.   In the Anglican church, younger ministers may expect to work in a junior capacity in two places for short periods before going on to be incumbents elsewhere for a longer period.   The pattern of moves may be constrained by diocesan boundaries, and often the number and the nature of moves early in the career may be related to the individual's eventual career prospects.   Williams, Blackstone and Metcalf (1974) have shown how the career structure of the university teaching profession affects the timing and direction of migration.

In teaching, McKay and Whitelaw (1977, pp. 39-40) found that mobility was very high in the State of Victoria, and that this was related to norms concerning career structure.   'Most teachers recognize that advancement in the profession depends upon a willingness, especially in the early years of a career, to transfer to other schools in

other parts of the state. Young teachers are
expected to serve their time in isolated rural
schools. However, as young families approach the
age when facilities for tertiary education are
required, there is a tendency for a return to metropolitan centers.' Zabalza (1978) discussed the
effects of differential salary structures between
districts on teacher migration in schools in
England and Wales.

Many professions are regulated by institutions
which control who is allowed to practice that profession. Sometimes their areas of jurisdiction
are small, and there may be barriers to movement
between them - for example, between states in the
USA. Here institutional barriers may take the
form of licensing or accreditation arrangements, and
the strength of the barriers will affect migration
in the relevant professions. Holen (1965) studied
the effects of state licensing policies on
interstate mobility in medicine, dentistry and law.
In medicine, there is a system of reciprocity in
existence, which normally allows doctors licensed
in one state to get a license when they move to
another. Such systems are weaker or nonexistent
in dentistry, where interstate migrants must take
an examination (with a high failure rate in some
states). Lawyers may or may not be required to
take an examination; their problem is the difference,
often very substantial, between the legal systems in
different states. As the above suggests,
physicians have a much higher interstate migration
rate than dentists or lawyers.

Advertising of Vacancies

The way in which job vacancies are advertised by
different organisations for different jobs is an
obvious but usually neglected factor affecting
migration flows. Although people may apply for,
and get, jobs which are never advertised, the types
of jobs which can be obtained in this way are
limited. Most changes of job result from an
employer letting it be known, selectively or
generally, that a job is to be filled, and potential
employees hearing of the job and enquiring about it.
From the employee's viewpoint, then, his or her
chances of hearing about a particular job depend
upon how it is advertised and whether he or she uses
the medium adopted.

The employer's choice of advertising medium

involves costs, and the extent of costs incurred
will depend on the employer's perception of what is
necessary to locate a suitable candidate.   If the
job concerned requires little skill or training,
there will be many suitable candidates and little
need to advertise it widely.   For a more
specialised job, it may be harder to find suitable
candidates and more costs will be incurred in
advertising the job widely.   Perhaps the cheapest
method of advertising a vacancy is word-of-mouth.
Only slightly more elaborate is the posting of a
notice at the entrance to the premises.   The use of
a labour exchange, a local newspaper, or a local
private employment agency are other ways of advertising vacancies at relatively low cost.   All
these methods are likely to reach a group of people
living very close to the place of employment, plus
others who are close friends or relatives of people
close to the workplace.   As a result, most jobs
of this type will be filled by local people, and
will not involve migration, although they may involve
occupational mobility.
     In contrast, jobs requiring more specialised
training, qualifications or aptitudes may need to
be advertised over a wider area.   This may involve
advertising in regional or national media, or in
information media which reach specialised groups of
people - special interest or trade journals are
obvious examples, as are the visits made by many
companies to universities in recruitment of management trainees.   Sometimes, particular journals
become known as media where certain kinds of jobs
are advertised; a British example is the role of
the Times Educational Supplement as an advertising
medium for school-teaching jobs.   Some companies
may wish to advertise for labour outside their own
areas, but may be able to economise by concentrating on a few areas where people with the required
skills are likely to be found, such as areas where
other companies in the same line of business are
located.   In multi-locational companies, internal
labour markets may exist, in which a job at one
place may be advertised at the company's other
establishments, more intensively than or to the
exclusion of advertising to the general public.
     If jobs are advertised outside the immediate
area, the chances of non-local people applying are
higher, and so presumably are the chances of the
appointee being non-local.   Jobs advertised over a
wide area are thus more likely to lead to migration.
As these jobs are disproportionately those in the

professional, managerial and technical socio-
economic groups, these groups could be expected to
have higher migration rates than those groups where
job advertising is predominantly local. This
effect can account for observed occupational
differentials in migration without recourse to ideas
that people in some occupational groups are more or
less psychologically tied to their home areas than
others.

Housing Institutions

The previous sections have discussed institutional
influences on job changes. In order for people to
take up job offers in other parts of the country,
however, they will normally need housing near their
new workplaces. Migration therefore is dependent
on the ability of such people to find housing which
is satisfactory to them. A major problem in this
respect is the difference between labour markets in
the price and availability of housing. It should
also be emphasised that there may be major
differences between housing sub-markets in a place,
perhaps with high-cost, high-value housing being
more easily available than low-cost housing.

The organisation of the housing market is
therefore an important factor in influencing inter-
urban as well as intra-urban migration. Ability
of owner-occupiers to move may be dependent on their
ability to find housing and to obtain a mortgage in
their new area. As such, they are dependent on
two sets of institutions, the real estate business
and the mortgage-granting institutions (such as
building societies in Britain and savings and loan
associations in the United States). The availability of mortgages may depend on institutional
regulations and perhaps on managerial discretion as
to whether mortgages will be granted to particular
people or on particular properties. Other insti-
tutionally-determined features, such as the interest
rate on mortgages, will affect the type of property
that a prospective owner-occupier can afford.

An individual household's access to housing may
depend on its search procedure, and particularly
the information sources used. The work of Palm
(1976) has indicated the spatial selectivity of
coverage of a real estate agent based in a particular location in a metropolitan area. Clark and
Smith (1979) have attempted to build this selectivity into a general model of the residential search

221

process, showing how the household's eventual destination may be affected by its use of the real
estate industry. Smith and Mertz (1980) have
extended this model by focusing on the strategies
real estate agents can adopt to influence the
eventual decision.

Methodological Problems of Institutional Research

The emphasis on institutional factors in explaining
internal migration patterns carries with it problems
concerning the restrictive nature of the generalisations which can be made. Conclusions must be dependent on the national context, because of their
dependence on the laws and constitution of the
country concerned. Even within one country, there
may be problems arising from structural or policy
differences between different types of employment.
Results from a study conducted in one context, then,
are not necessarily capable of extension to another
context. This is a major difficulty in constructing
general theory about institutional effects on
migration.

There are practical problems in conducting
research on the effects of certain kinds of institutional influences on migration. A study of organisational influences on employee migration, for
example, may encounter problems from companies who
make such decisions on an ad hoc basis without ever
formulating a policy. In other cases, a policy
may exist but may be carried out inconsistently or
not at all. Records on employee migration may not
exist or, if they do exist, may be confidential and
unavailable to the researcher. Both policy statements and records are likely to use terminology
specific to the company concerned, making comparisons
difficult; in addition, there may well be differences
between branches of a company, and frequent organisational and policy changes through time.

The general problem that policies in use may
differ from stated policies creates many methodological difficulties for the institutional researcher,
whether such differences are due to local discretion
or to a reluctance to admit discreditable, controversial, or chaotic aspects of policy in use.
These differences may be hard to identify and, when
identified, to substantiate, especially when the
institution concerned is unwilling to admit their
existence.

Conclusion

There are many respects in which an institutional approach can offer new insights into the process of internal migration. It does not, however, offer as neat or coherent a view of the process as do economic, behavioural or spatial models of migration. Even though institutional factors make it harder to formulate general theory, it is important that they are taken into account, both because they are directly responsible for a large number of individual moves and because they are necessary to account for the failures of more general models to fit a given data set.

At the current state of development of institutional approaches in geography, explanation tends to be highly specific to individual cases. Because of the dependence of findings on specific institutional situations, these ad hoc explanations may never be fully integrated in a more general theory, but it is hoped that a greater degree of generalisation will be attainable eventually. The best approach to this goal may be through development of closer links to theories of institutional behaviour. McKay and Whitelaw (1981) have probably made most progress in this direction in their study of executive mobility in large organisations, especially in their attempts to link organisational mobility policies to management goals in executive career development. However, it should be anticipated that institutional theories of migration will be based only on a typology of institutional activity, and that reliable predictions of specific institutional effects on migration will be difficult or impossible.

The main disadvantage, then, of an institutional approach is that it lacks generality and power to make specific predictions. Against this, however, it offers a convenient way of combining insights from different disciplinary strands of research on migration. It also focuses on variables that are amenable to organisational or governmental control, and may therefore have direct policy relevance. Finally, it is less abstract and far more intuitively reasonable than most attempts at creating a general theory of migration.

Robin Flowerdew

References

Barsby, S. and Cox, D.R. (1975) Interstate Migration
    of the Elderly: An Economic Analysis, Lexington
    Books, Lexington, Massachusetts
Bateman, M., Burtenshaw, D. and Hall, R.K. (1971)
    Office Staff on the Move, Research Paper No. 6,
    Location of Offices Bureau, London
Beaumont, P.B. (1976a) 'The Problem of Return
    Migration under a Policy of Assisted Labour
    Mobility: An Examination of some British
    Evidence', British Journal of Industrial
    Relations, 14, 82-8
Beaumont, P.B. (1976b) 'Assisted Labour Mobility
    Policy in Scotland 1973-1974', Urban Studies,
    13, 75-9
Birch, S. and Macmillan, B. (1971) Managers on the
    Move: A Study of British Managerial Mobility,
    Management Survey Report No. 7, British
    Institute of Management, London
Cebula, R.J. (1974) 'Interstate Migration and the
    Tiebout Hypothesis: An Analysis According to
    Race, Sex and Age', Journal of the American
    Statistical Association, 69, 876-9
Cebula, R.J. (1979) The Determinants of Human
    Migration, Lexington Books, Lexington ,
    Massachusetts
Clark, G.L. (1982) 'Volatility in the Geographical
    Structure of Short-run US Interstate Migration',
    Environment and Planning A, 14, 145-67
Clark, W.A.V. and Smith, T.R. (1979) 'Modeling
    Information Use in a Spatial Context', Annals
    of the Association of American Geographers, 69,
    575-88
Desbarats, J.M. (1977) 'Estimating External Constr-
    aints to Migration', Professional Geographer,
    29, 283-9
Flowerdew, R. and Salt, J. (1979) 'Migration between
    Labour Market Areas in Great Britain, 1970-1971',
    Regional Studies, 13, 211-31
Gleave, D. and Cordey-Hayes, M. (1977) 'Migration
    Dynamics and Labour Market Turnover', Progress
    in Planning, 8, 1-95
Gleave, D. and Palmer, D. (1980) 'The Relationship
    between Geographic and Occupational Mobility in
    the Context of Regional Economic Growth' in
    J. Hobcraft and P.H. Rees (eds.), Regional
    Demographic Development, Croom Helm, London,
    pp. 188-210
Gober, P. (1981) 'Interregional Migration and Federal

Policy in the U.S.', paper presented at the Annual Meeting of the Association of American Geographers, Los Angeles

Greenwood, M.J. and Sweetland, D. (1972) 'The Determinants of Migration between Standard Metropolitan Statistical Areas', Demography, 9, 665-81

Haber, S.E. (1981) 'The Mobility of Professional Workers and Fair Hiring', Industrial and Labor Relations Review, 34, 257-64

Holen, A.S. (1965) 'Effects of Professional Licensing Arrangements on Interstate Labor Mobility and Resource Allocation', Journal of Political Economy, 73, 492-8

Johnson, J.H. and Salt, J. (1980a) 'Employment Transfer Policies in Great Britain', Three Banks Review, 126, 15-39

Johnson, J.H. and Salt, J. (1980b) 'Labour Migration within Organizations: An Introductory Study', Tijdschrift voor Economische en Sociale Geografie, 71, 277-84

Johnson, J.H. and Salt, J. (1981) 'Population Re-distribution Policies in Great Britain' in J.W. Webb, A. Naukkarinen and L.A. Kosinski (eds.), Policies of Population Redistribution, Geographical Society of Northern Finland, Oulu, Finland, pp. 77-92

Johnson, J.H., Salt, J. and Wood, P.A. (1974) Housing and the Migration of Labour in England and Wales, Saxon House, Farnborough, Hampshire

Klaassen, L.H. and Drewe, P. (1973) Migration Policy in Europe: A Comparative Study, Saxon House, Farnborough, Hampshire

Ladinsky, J. (1967a) 'Sources of Geographic Mobility among Professional Workers: A Multi-variate Analysis', Demography, 4, 293-309

Ladinsky, J. (1967b) 'Occupational Determinants of Geographic Mobility among Professional Workers', American Sociological Review, 32, 253-64

Leslie, G.R. and Richardson, A.H. (1961) 'Life-cycle, Career Pattern and Decision to Move', American Sociological Review, 26, 894-902

Lewis, R.A. and Rowland, R.H. (1979) Population Redistribution in the USSR: Its Impact on Society, 1897-1977, Praeger, New York

Long, L.H. and Hansen, K.A. (1979) Reasons for Interstate Migration: Jobs, Retirement, Climate, and Other Influences, Current Popula-tion Reports, Series P-23, No.81, US Bureau of the Census, Washington,DC

Robin Flowerdew

Mann, M. (1973) <u>Workers on the Move: The Sociology of Relocation</u>, Cambridge University Press, Cambridge

McKay, J. and Whitelaw, J.S. (1977) 'The Role of Large Private and Governmental Organizations in Generating Flows of Inter-regional Migrants: The Case of Australia', <u>Economic Geography</u>, <u>53</u>, 28-44

McKay, J. and Whitelaw, J.S. (1981) 'Organisations, Management and Structural Change: The Role of Executive Mobility' in G.J.R. Linge and J. McKay (eds.), <u>Structural Change in Australia: Some Spatial and Organisational Responses</u>, Department of Human Geography, Australian National University, Canberra, pp. 87-118

Meinig, D.W. (1965) 'The Mormon Culture Region: Strategies and Patterns in the Geography of the American West, 1847-1964', <u>Annals of the Association of American Geographers</u>, <u>55</u>, 191-220

Öberg, S. and Oscarsson, G. (1979) 'Regional Policy and Interregional Migration - Matching Jobs and Individuals on Local Labour Markets', <u>Regional Studies</u>, <u>13</u>, 1-14

Ostrosky, A. (1978) 'Some Economic Effects and Causes of State and Local Government Commitment to Public Education', <u>Indian Journal of Economics</u>, <u>60</u>, 49-58

Palm, R. (1976) <u>Urban Social Geography from the Perspective of the Real Estate Salesman: Minneapolis and San Francisco Case Studies</u>, Center for Real Estate and Urban Economics, University of California, Berkeley, California

Rapoport, R. and Rapoport, R.N. (1971) <u>Dual-career Families</u>, Penguin Books, Harmondsworth, Middlesex

Roseman, C.C. (1977) <u>Changing Migration Patterns within the United States</u>, Resource Papers for College Geography No. 77-2, Association of American Geographers, Washington, DC

Scipione, P.A. (1973) 'A Computer Solution for Determining Student Migration', <u>Professional Geographer</u>, <u>25</u>, 249-54

Smith, T.R. and Mertz, F. (1980) 'An Analysis of the Effects of Information Revision on the Outcome of Housing-market Search, with Special Reference to the Influence of Realty Agents', <u>Environment and Planning A</u>, <u>12</u>, 155-74

Taylor, R.C. (1969) 'Migration and Motivation: A Study in Determinants and Types' in J.A. Jackson (ed.), <u>Migration</u>, Cambridge University Press, Cambridge, pp. 99-133
226

US President's Commission for a National Agenda for
     the Eighties (1980), A National Agenda for the
     Eighties, US Government Printing Office,
     Washington, DC
Williams, G., Blackstone, T. and Metcalf, D. (1974)
     The Academic Labour Market: Economic and
     Social Aspects of a Profession, Elsevier,
     Amsterdam
Wiltshire, R. (1979) 'Research on Reverse Migration
     in Japan: (II) Personnel Transfers', Science
     Reports of the Tohoku University, 7th Series
     (Geography), 29, 135-42
Zabalza, A. (1978) 'Internal Labour Mobility and
     the Teaching Profession', Economic Journal, 88,
     314-30

Chapter 8

RESIDENTIAL MORTGAGE PATTERNS AND INSTITUTIONAL STRUCTURES: THEORY AND REALITY

Peter Williams

Introduction

Urban geography has long been concerned with
patterns of residential location and models of urban
structure (Johnston, 1980).   Despite agreement over
certain observable regularities in the spatial distribution of social classes and life-cycle groups
there continues to be considerable debate over the
explanation of this empirical evidence.   Indeed,
within urban geography we have witnessed the use of
Darwinian ecological models and their derivatives
through factorial ecology, behavioural and inter-
actionist approaches, trade-off models deriving from
neoclassical economics, Weberian urban managerialism
and most recently Marxist political economy as
frameworks for the explanation of spatial distributions.   Each approach has asserted the importance
of different relationships and processes while the
level and complexity of the concepts used varies
from explanations at the level of individual
behaviour to an analysis of the nature of capitalism.
The purpose of this chapter is to consider the links
between explanation at the level of societal processes and the operational behaviour observable in
the allocation of residential mortgages at the
branch level.   As such, the chapter seeks to draw
together the concerns of urban political economy
and urban managerialism to provide a framework for
the interpretation of urban phenomena at the local
level.   It is perhaps worth stressing at this
point that the term 'urban' is used simply for con-
venience; it is not intended to imply that a set of
distinctive urban processes can be identified.

Peter Williams

Institutions and Structures

Institutional urban research in the United Kingdom
drew its conceptual underpinnings from the work of
Pahl and the so-called urban managerialism (Pahl,
1970).   In seeking to reconstruct urban sociology
Pahl proposed a sociology of access and allocation
based around the study of the distribution of goods
and services in urban areas and the role of particular officials who, by reason of their position
within particular institutions and agencies, were
able to influence significantly the outcome of
allocational procedures.   This perspective stimulated a number of studies of planners, housing
officers, building society managers and estate
agents (see Bassett and Short, 1980a), each of which
gave empirical support to the view that these institutions or organisations did play an important role
in determining the form and structure of urban areas
in general and residential patterns in particular.
     However, the managerialist perspective had pro-
found weaknesses.   There was no clear framework
within which to locate and interpret managerial
activity, nor did there appear to be any prescribed
limits on the power of individual managers, thus
giving grounds for the critique of managerialism
which developed (Gray, 1976).   Pahl himself sought
to reformulate the concepts, but in the context of
a general upsurge in interest in Marxist political
economy, research related to specific managerial
activities became less relevant.   The institutions
within which such managers were located were
inserted into logically structured accounts of the
nature of capitalism (e.g. Boddy, 1976a) which gave
clear insights into why resources were scarce.   The
question of discretion was necessarily diminished
when set against a context of overall scarcity and
inequality under capitalism.
     The development of a political economy of
urbanism has been of vital importance in the
evolution of urban studies.   As might be expected,
however, there are weaknesses in the approaches
adopted.   In particular, it is apparent in retrospect that much of the work carried out within this
rubric was crudely functionalist, implying as it did
that the needs of capital were so dominant that all
phenomena could be explained by them.   Moreover,
this rendered all forms of opposition, whether
working-class or from within the capitalist class,
largely irrelevant, except perhaps when some

230

concession was obtained which less obviously
supported dominant interests.   Finally economic
relationships were given supreme power with ideology
and politics being reduced to secondary factors.
Increasingly the failings of such accounts are being
recognised and attention is being directed to respecifying abstract and highly aggregate concepts in
ways which will assist a much more careful and sensitive analysis of social realities.   Furthermore,
within this, the place of empirical work has been re-
established in recognition that there are many gaps
in our knowledge of both the structures and processes in urban areas.

This reappraisal of Marxist urban research
provides an opportunity for reconsidering the place
of institutions in urban analysis.   While the
weaknesses of the managerial approach are readily
conceded, the empirical focus on managers and institutions still has much validity.   As Pahl (1979,
p. 89) stated, 'whatever theoretical stance one
adopts in relation to class in capitalist society
independent analysis of the instruments and agents
of allocation is still necessary.'   To grasp the
processes which structure daily life an understanding of the mediating role of institutions and
agencies is essential.   After all, to reflect upon
structure is not simply to think of abstract processes separate from social realities.   It is also
to consider how those processes came to take the
form they do, how they are shaped and transformed
within our society.   Thus, although one might con-
cede the importance of the process of accumulation,
the search for profit or the role of class struggle,
at some point and in some way these take material
form.   Analysis at the level of institutions and
agencies provides one such opportunity to consider
this.

Conceiving of urban geography as being concerned with those social processes, including the
role of space, which find expression in the structuring of urban areas then there is clearly a place
for considering how institutions such as housing
departments, estate agents or financial agencies
actually mediate those processes.   Part of the
debate will revolve around the question of appropriate institutions, an issue compounded by conventional perspectives on the value of the 'urban'.
The identification of particular managers (e.g.
building society branch managers) or particular
institutions or agencies (e.g. planning departments)
has led to some confusion.   Are we concerned with

managers or institutions and is a planning department
an institution or part of one, i.e. the local
authority?   These questions are of some importance
because they mark out a sphere of interest and an
empirical starting point.   To separate managers
from institutions is to pull individuals out of their
context.   As will be argued later this would be unhelpful and reinforce the possibilities of crediting
too much independence to managers.   To that end it
is probably correct to use the term institutional
rather than managerial.   The same problem of
independence arises with respect to a single department within a local authority.   However, assuming
the department is of some significance with an
identifiable budget and function the grounds for
treating it as a starting point are much stronger.
Moreover, to ignore the divisions within the
apparatuses of government at either central or local
level is to fundamentally misunderstand the nature
of the State.   In this chapter the terms institution, organisation and agency are used inter-
changeably.   It is also appropriate to note that
when managerial roles are described these are always
seen in the context of the agency concerned, i.e.
they are not independent.
     Elsewhere I have argued that even accepting a
basic Marxist framework we are still left in a
dilemma when it comes to tackling questions regarding the role of institutions in the allocation
process (Williams, 1982).   This is because many of
the concepts of Marxist analysis have not been
specified in sufficient detail to assist the precise
examination of social process.   To that end the
developing literature on the political economy of
organisations is helpful (e.g. Clegg and Dunkerley,
1980) as are attempts to specify more precisely
different levels of abstraction (Gibson, 1981)
and to consider the relationships between structure
and agency (Gregory, 1981).   Such debates raise
important questions with respect to analysis of
organisations and institutional procedures and take
us part of the way in understanding how market processes ultimately find their expression upon the
ground.   In the second part of this chapter we
undertake an examination of one specific set of
institutions in their market setting - the building
societies and their role in the allocation of
mortgage funds.

Mortgage Allocation

The market for mortgages in Britain is dominated by
the building societies (Table 8.1). This position
has been achieved over several decades of fairly
intense competition with banks, insurance companies,
solicitors and local authorities. The societies'
position now seems relatively secure and, with continued government support, they are likely to
retain this position for the foreseeable future.
This should not, however, lead to assumptions that
the societies are not changing or that their position
with respect to the finance market will not change.
Much of this is beyond their direct control and
arises as a consequence of changes within the
economy as a whole.

In Britain, there has been considerable interest within urban research in the role of building
societies. In numerous accounts their importance
in the allocation of mortgages has been noted (e.g.
Lambert, 1976), pointing towards the ways their
rules and priorities dictate the structure of the
housing market. Given that data on mortgaging has
been relatively easy to obtain, especially at an
aggregate level, the links between mortgage allocation and social class and mortgage allocation and
property type have been well established (Duncan,
1976; Boddy, 1980). Other issues have been uncovered, such as the societies' reluctance to
advance funds on non-standard property, their conservatism, and their role in redlining areas
(Williams, 1978). These have perhaps been the key
concerns. There have also been limited attempts
to construct a political economy of the local
'market', linking agents, solicitors, property companies and societies into an interconnected web,
pursuing their own self-interest (Boddy, 1976a;
Community Development Project, 1976). Of work
elsewhere, research in the United States on mortgaging questions has been quite substantial being
particularly directed to the issues of redlining and
gentrification (Smith, 1979; London, Bradley and
Hudson, 1980; Tomer, 1980). Moreover, Harvey in
his research on Baltimore has probably constructed
the most substantial account of the interrelation-
ships between the financial sector and urban housing
markets. In a series of publications (1973; 1974;
1975; 1977; 1978) he interprets the role of mortgage
institutions as part of the growing dominance of
finance capital over industrial capital and relates

233

the general process of structuring urban areas to consumption patterns and the fragmentation of classes.

The underlying perspective for work in Britain has ranged from managerial to Marxist. In part, inspired by Harvey, there has been a tendency to develop, from the detailed evidence already referred to, certain generalisations which fit within a Marxist framework (Boddy, 1976a; 1976b; Bassett and Short, 1980b). Thus the patterns of allocation have come to be seen as logical con- sequences of market relationships. Redlining is related to central city disinvestment and suburban reinvestment. Managers far from having discretion are seen to act in accordance with priorities handed down from the logic of the economy. These assessments relate closely to accounts which locate building society operations within the circuits of capital and the processes of commodity production and consumption in the housing sphere.

This work has been of major importance but as has been suggested earlier it has led to a diminu- tion of interest in the processes of allocation and close consideration of operations at the local level. It has also given rise to generalisations which are to say the least somewhat crude, e.g. the implication that societies in following the same logic must therefore perform in similar ways, or that branch managers have no discretion. My con- cern is not with the fact that these generalisations are made, since it is necessary to extract from detailed information certain key positions. Rather it is that within the frameworks laid out by contributors such as Harvey (1977) or Johnston (1980) there seems to be little place for approach- ing questions related to historical change, organ- isational processes, and institutional mediation. Admittedly this is to press for a highly disaggre- gated conceptual framework but that is what is required if we are to approach the complexity of urban change over time. In the following section we develop this argument through a selective exam- ination of building society operations in Britain.

Building Societies and Mortgage Allocation: Empirical Examples

In this section of the chapter we are concerned with relating some of the earlier argument to sel- ected aspects of building society activity. It is

Table 8.1: Annual Net Amount Loaned for House Purchase (Total Loaned − Amount Repaid) by Building Societies, Insurance Companies, Local Authorities and Banks, 1959-78, UK

Year	Building societies £m	%	Insurance companies £m	%	Local authorities[b] £m	%	Banks £m	%	Total[a] £m
1959	231	64	35	10	26	7	70	19	362
1960	240	63	68	18	42	11	30	8	380
1961	221	55	81	20	67	17	30	8	399
1962	276	67	61	15	47	11	30	7	414
1963	422	74	34	6	59	10	55	10	570
1964	546	72	53	7	121	16	35	5	755
1965	459	65	91	13	168	24	-15	-2	703
1966	667	88	61	8	53	7	-25	-3	756
1967	823	85	34	4	68	7	40	4	965
1968	860	89	71	7	9	1	25	3	965
1969	782	93	83	10	-18	-2	-5	-1	842
1970	1088	88	36	3	72	6	40	3	1236
1971	1600	92	13	1	107	6	90	5	1736
1972	2215	80	2	-	198	7	345	12	2761
1973	1999	72	121	4	355	13	310	11	2785
1974	1490	68	120	5	559	23	90	4	2202
1975	2768	81	67	2	620	16	60	2	3428
1976	3618	97	13	-	67	1	70	2	3737
1977	4100	96	22	1	5	-1	120	4	3921
1978	5096	95	47	1	-57	-1	270	5	5356

Note: a. Excludes advances by new towns and the Housing Corporation, which totalled: 1971, £12 million; 1972, £22 million; 1973, £46 million; 1974, £113 million; 1975, £135 million; 1976, £105 million; 1977, £28 million; 1976, £-34 million.
b. Includes loans for improvement etc., housing associations, and sale of council houses.

Source: Housing and Construction Statistics (HMSO). (Taken from Boddy (1980)).

not possible in a single chapter to cover in detail
the many issues which require elucidation. Furthermore, evidence on some features is presently unavailable and all we can do is indicate areas of
importance. This exploration begins with a brief
comment on the development of the building society
industry in the context of the financial market as
a whole and secondly with regard to relations within
the industry. Subsequently discussion focuses on
operations in a local housing market.

There currently exist a number of studies
which partly explain the evolution of building
societies in relation to the financial market
(Bellman, 1927; Boddy, 1980; Cleary, 1965; Gauldie,
1974; Pawley, 1978). However, we remain unclear as
to the process by which societies became an
essential component of the market for interest-
bearing or credit capital. We can identify their
function in terms of transforming money into
capital to assist the realisation of value in newly
built dwellings and as part of a process of extract-
ing value, obtained via the wage, through charging
interest on loans for house purchase. Those relationships not only reflect the position of housing
as both a necessity and as a commodity but also the
evolution of the finance market in a way which
allowed societies to become established and move
into a position of dominance in the home loans
market. Despite the somewhat eulogistic accounts
from the building society industry itself, which
generally take the form of statements about working
men gathering together to house themselves, it does
appear that even the earliest terminating societies
were predominantly organised by the working and
lower-middle classes (Godsden, 1974). They were
not only a means of obtaining housing but were also
closely linked to the ideology of self-help and
thrift and a means for breaking the loan finance
monopoly held by solicitors and earlier by scrive-
ners (brokers). But, as the Royal Commission on
Friendly and Benefit Building Societies (1872) noted,
by the time permanent societies were established in
number, the societies had very definitely become
middle-class preserves with their managers being
drawn from the professional and small merchant
classes, a point also made by Engels (1970).

Not only did the societies provide much needed
funds, which might otherwise have been unobtainable
given the state of the banking system, but they also
created a means for individuals to buy and construct
homes either for their own use or to rent. The

236

mobilisation of savings and other investment money
provided an important route for the expansion of
the housing market. Many were interested in this
including landowners, builders and exchange professionals (estate agents, solicitors, brokers) as
indeed they are today. There was also, prior to
1871, substantial interest by the political parties
since property ownership was a means of enfranchis-
ing the populace.

     The establishment of societies, and their
development within the sphere of credit, set in
motion a substantial motor force with respect to
the housing market, a process which ultimately was
to assist the decline of the private rented sector
and the growth of home ownership. As investment
alternatives grew through a more developed financial
system private mortgages through solicitors were
harder to obtain. Moreover, many of these
exchange professionals were now involved in the
management of the societies, giving them
opportunities to feed funds into home ownership with
all its attendant impact on conveyancing fees,
market turnover and price inflation. The instability of the private rented market contrasted with
the apparent security of home ownership and this
encouraged societies to cultivate that sector (Kemp,
1980). As competition in the financial market
grew the societies focused their efforts on a major
growth area - the emergent home-owning class.
Throughout this century, despite conflicts within
the mortgage industry, the societies have extended
their control over this sphere. Indeed, with substantial government support, they have been able
to dominate the market for personal savings. In
1966 building societies' assets were £6,350 million
and banks were holding personal deposits of £7,466
million. By 1978 the societies' assets were £39,692
million and bank deposits stood at £24,490 million.

     This gives some indication of their success.
Although the banks have consistently criticised the
building societies' privileged position (see, for
instance,the Committee to Review the Functioning of
Financial Institutions, 1980), their growth has
gone unchallenged from many quarters. The state
has generally supported the industry, in part
because it helped meet housing objectives. Manufacturing and commercial interests were perhaps
more ambivalent because of fears related to the
impact of home ownership on labour mobility. In
recent years these pressures have increased.
Industrialists are just one of a number of factions

who have criticised the societies and asked for a reduction in their dominance. As the societies have come to be important within the financial market as a whole, influencing market rates of interest and flows of capital as a consequence of their size, so antagonism has increased (Rose, 1978).

These forces have acted to structure relationships within the industry. Although many of the early societies were terminating, and thus did not compete in the formal sense with each other, there was much interest in their activities because of the impact of building on local social relations. Builders, landowners and solicitors all benefited from the transactions which were generated - just as they do now. With the establishment of permanent societies, the mutual aspects of their operations were increasingly undermined and the distinction between a society as an institution or organisation and its membership grew. Paid officials became more commonplace and the volume of transactions a society could sustain rose. As the opportunities for profit in the housing market expanded so the competition between societies increased. Indeed, the formation of societies became in itself a profitable activity, as instanced by James Starr, who franchised a whole series of Starr-Bowkett societies in the middle of the late nineteenth century, deriving his earnings from (Godsden, 1974, p. 169)

> the fees he charged from the secretary, solicitor and surveyor on their appointment to one of his societies. To ensure they in turn were able to reap their necessary rewards the rules were drawn in such a way as to make the dismissal of these office-holders by the members virtually impossible.

It is often argued that because building societies are not profit-making they have no incentive to expand except through the selfish desires of their own management to enhance their salaries. There is certainly evidence to suggest expansion has been related to salary rewards (Gough, 1979) but that is too limited an explanation to suffice. Societies compete with each other and with the other institutions in order to survive. A society that ceases to expand will decline firstly in relative terms and probably absolutely. In part, this is because there is little to distinguish

between the services of one society or another and
thus any society which loses its momentum as an
enterprise will become increasingly marginal.

The establishment of the Building Societies
Association in 1936 and the institution of cartel-
like arrangements with respect to the fixing of
share, deposit and mortgage rates reduced the possibilities of competition between societies and acted
to marshall the societies' efforts to expand the
overall market share held by the industry. At the
same time, as Gough and Taylor argue (1979, p. 48),
'The cartel in interest rates has inevitably led
societies to compete in other ways. The principal
ones are product differentiation, advertising,
branch expansion and growth via merger.' In the
late 1970s, there were signs of the cartel breaking
down as a number of large societies chose to move
away from the agreed rates on deposits, and the
latest evidence is that the cartel will disappear
(Williams, 1981). This change reflects competitive
pressures in the savings market. Some societies
are more efficient than others and are able to
function with lower management costs. The difference between the rate of interest paid on investments and that on mortgages can thus be narrowed by
changing one of the rates. Since the demand for
mortgages is excessive and the charge for loans is
believed to be 'acceptable' the societies have
generally chosen to increase the rate paid on investments.

The competitive consequences of this decision
are enormous and are part of an interest-rate spiral
which has affected government programmes and industrial activity. This is simply one element in a
general deepening of competition between societies
and within the market in general. For example, the
use of advertising is now substantial, reflecting
the drive by all institutions to capture a larger
share of the market for personal savings; in 1979,
the societies spent over £14 million, the Abbey
National spending £2,309,000 and the Halifax
£2,337,600. Indeed, the societies are now some of
the largest spenders on television advertising (as
they are on office refurbishment, computer technology
and other areas). There is also growing competition in investment strategies with the introduction
of term shares, savings bonds, linked savings
accounts and other devices. In addition societies
have increased their activity in the areas of
sponsorship, housing policy initiatives and most
recently moves to establish European links. These

strategies must also be viewed in the context of
massive branch and staff expansion programmes and a
whole series of mergers between societies.   In
other words, building society operations have been
restructured to meet the new phases of competition.
    Between 1960 and 1978, the number of societies
fell from 726 to 339.   At the same time, the number
of branches increased from 900 to 4,411, thus
making the societies the fastest-growing sector in
the retail shop market.   The expansion reflects the
belief that it is necessary to have highly accessible outlets for depositors, particularly in
shopping centres and areas of high pedestrian
traffic, thus creating 'money' shops.   The
strategy is correct in that branch openings can be
seen to produce revenue and to generate interest in
the societies concerned (Davies and Davies, 1979).
However, local authorities and retailing associations
are increasingly resisting the opening of more
branches, as some streets are now swamped with
building society offices (Woolwich Review, 1981)
thus diminishing passing trade and, ultimately,
undermining the societies themselves.   Furthermore,
societies have found that when mergers are arranged
overlapping branch networks are now a substantial
problem.   Thus when the Anglia Building Society and
the Hastings and Thanet Building Society merged, out
of the 194 locations in which there were branches,
18 overlapped.
    Mergers are now very common in the industry.
The search for the size necessary to sustain growth
and profits and to be able to use technology
efficiently has pressured many societies to seek
partners.   In 1978, 22 such mergers took place.
Both large and small societies have been involved in
this process with encouragement from the Chief
Registrar who is the government official responsible
for administering the Building Societies Act.   He
has taken the view that small societies are vulnerable, a view which the evidence supports.   The
problems of the Wakefield, Grays and Chesham
Building Societies in the 1970s can all in part be
explained by the problems of developing an adequate
management structure in a small enterprise (Registry
of Friendly Societies, 1979).   The Registrar
clearly recognises that support continues for small
societies (though it is interesting to ask from
whom) but that the future of the industry will increasingly be in the hands of five or ten major
societies.   At present five societies, Halifax,
Abbey National, Nationwide, Leeds Permanent, and

Woolwich, have over 55 per cent of the industry's
assets and, as the annual reports show, this
importance has steadily grown. In 1930, they only
held 39.1 per cent of assets.

The intensity of competitive pressures within
the industry is often hidden from view, though it
surfaces periodically - for instance in the splits
within the industry in the 1930s leading up to the
1939 Building Societies Act (Cleary, 1965) and in
1956 when the Halifax left the Building Societies
Association over the question of regulations on the
amount of company funds entering the societies.
Such public splits are less likely to occur now
because the societies have sought to build the
Building Societies Association into a substantial
guard between themselves and the government.   To
erode its authority publicly would be to open up a
wide range of questions during a particularly sensitive period.

The structure of competition both outside and
within the building society industry shapes operations at the local level.   Competitive pressures
ensure that branch managers must act primarily as
investment getters, giving housing a secondary
position.   More to the point, it would seem that
the position of societies in the investment market
influences lending behaviour.   Thus the Abbey
National has been able to initiate developments in
the housing sphere partly because it has established
such a strong investment base.   Moreover, probably
the most conservative societies in terms of lending
policy are those in the middle range which neither
rely solely on a local base nor have national status
to maintain their market share; in a sense they are
doubly vulnerable.   As such they have an aggressive
stance in the market place.   They are more likely
to rely on 'hot' money, i.e. funds which are
switched according to the highest rates of interest
available.   They also tend to be restrictive in
terms of the percentage advance they will grant and
the types of properties they are prepared to lend
on.

Through study of these practices, we are able
to gain some insights into the relationship between
broader structures and detailed aspects of policy.
The logic of the societies' behaviour is easily
understood, yet it would be wrong to assume that all
relationships follow such logical paths and in subsequent paragraphs we consider operations at the
local level in the light of the discussion so far.
In particular we are concerned with teasing out

241

questions of variation in local operations.

Institutional Structures and Competition in the Urban Environment

The changing form and structure of the building
society industry at national level is also observable
in local operations.   In this section empirical data
are drawn from a study of building society activity
in Dudley, West Midlands (Housing Monitoring Team,
forthcoming), supplemented by additional research on
this locality.

The town of Dudley and indeed the Black Country
as a whole was one of the centres of development of
the building society industry in the eighteenth and
nineteenth centuries.   One of the first recorded
societies was the Dudley Arms Building Society
(1786) whilst the still-functioning Dudley and
Rowley Regis Building Societies were formed in 1789
and 1792 respectively.   By 1930, the level of
activity in the town had consolidated around the
operations of six societies.   All had strong local
followings and were largely controlled by local
entrepreneurs and professionals. Until the late
1950s these societies dominated local lending and
they grew, as did the level of owner-occupation.
However, their dominance was increasingly challenged
by other societies which were now expanding their
branch networks.   This process of expansion usually
began with the opening of agency arrangements in
Dudley.   In 1964, the Halifax became the first
national society to open a branch there and from
that point on the picture changed rapidly.   In 1960,
there were six offices of the local societies and
twenty-three agencies.   By 1970, there were sixteen
branch offices and fifty-three agencies and in 1979,
thirty-six branches and forty-five agencies.

What we are observing here is a major restructuring of local credit operations.   The local
societies, controlled by members of local elites,
were progressively undermined by incoming societies.
The local societies failed to sustain their share of
lending and one by one were taken over by other
societies operating within the region.   The Sedgley
and District Permanent transferred engagements to
the South Staffordshire in 1959, the Halesowen
Benefit to the Redditch and Worcester in 1968, the
Stourbridge Lye and District Permanent to the
Coventry Economic in 1976, the Hasbury and Cradley
District Benefit to the Midshires in 1978, and in
1979 the Brierley Hill and Stourbridge Incorporated

merged with the Leamington Spa. Thus by 1979 the
only remaining local society was the Dudley Building
Society. Interestingly all the societies listed
were absorbed by regional rather than national
societies. The explanation for this is that the
former were struggling to develop an identity and
were working on the belief that a regional identity
was the strongest base; the latter had an established
reputation and by now were expanding via their own
branch development programme. Thus in Dudley, local
societies disappeared to become branches of aspiring
regional societies, national societies opened their
own branches and by the later 1970s agency operations
were in decline.

In all likelihood this restructuring has had
important implications for the local market though
it is not possible to quantify these effects. It
is reasonable to suggest that a full mobilisation of
local savings might not have occurred if these
changes had not taken place. It is also the case
that, being inserted into a national network of
societies, Dudley could attract funds which were not
generated there. Those inter-regional transfers
are one way of sustaining and developing a market in
the absence of a local economy capable of producing
the returns necessary for its own growth. Perhaps
the most obvious impact of the changes recorded is
upon direct local control. The opportunities for
local solicitors, accountants, builders and estate
agents to sit on the boards of societies has diminished, thus reducing the ready and quick access to
funding which once was available. Of course, they
can still develop a relationship with the branches
which are now more prevalent in the town but the
rules are no longer of their own making and an
important element of control has slipped from their
grasp. The 'local boards' which some societies,
having taken over a local society, erect to lay
claim to local loyalties are no substitute here.

This is not to diminish the impact of the far
greater number of societies now active in the area.
The branch managers find themselves in a competitive
environment in which the search for funds is para-
mount. Many societies insist on their managers
spending a large portion of each day making the
rounds of their area drumming up business. This
includes visits to accountants, solicitors, brokers,
estate agents, local firms (for the employees) and
local schools (for their children). This work is
closely monitored, with societies being able to cost
the effectiveness of particular types of visits. The

pressure on local managers to produce funds is quite intense (though it varies substantially between societies) - their own prospects for promotion being partly tied to their success in the investment field. Since societies offered relatively little variation in terms of investment opportunities, managers have to ensure that they maintain frequent and close contact and that mortgage funds - one of their few bargaining cards - are kept available. If the estimate given by Boddy (1980) that up to 40 per cent of funds are generated through professional contacts is correct, the importance of this work should clearly not be understated.

The expansion of branch networks has not only led to greater competition but has led to a major growth in recruitment of society staff. In 1969 building societies as a whole employed 23,549 staff. By 1978 this figure had risen to 44,932 employees. In recent years branch managers have found that pro- motion opportunities were considerable. Movement between branches was rapid, reducing to some extent the level of knowledge and involvement a manager might have of a particular locality, and ensuring that the local network was constantly disrupted by the arrival and departure of personnel. Again, it is difficult to quantify the significance of this aspect of the development of credit structure. However, it will have acted to disrupt the comfort- able informal networks built up over a long period of time and to substitute more formal market criteria. Thus relationships between market pro- fessionals and managers will be mediated through flows of finance (investments and mortgages) rather than less measurable networks of favours. Finally, an emerging issue of some importance in relation to staffing and staff morale relates to mergers and the possibilities of diminished expansion. It seems quite likely that the building society industry will soon enter a phase where promotion opportunities will begin to contract and the societies will then have to adopt new strategies to maintain the motiva- tion of their staff and savings inflows.

The changing structure of the industry also impinges upon estate agents, solicitors and others. Many of these exchange professionals had previously had building society agencies. Such agency arrange- ments not only provided useful revenue for the up- keep of premises but also ensured ready access to funds. When the society concerned opens up a branch, the agency arrangement is normally terminated and the firm involved is compensated. As the figures

for Dudley show, agency arrangements are in decline -
societies preferring the direct control which they
obtain through a branch. Local professionals have
thus not only lost direct control of local societies
but also access via agency arrangements which are
now increasingly difficult to obtain. At first
glance the structure of interests operating around a
local market might appear stable - what we are
trying to show here is that a considerable degree of
change occurs and that underlying the calm of a local
market are a series of significant conflicts as
societies restructure their operations, individual
managers continue to search for investment and local
professionals seek the funds necessary to maintain
the turnover of property and buoyancy of market
conditions.

Lending Practices: Discretion or Constraint

Consideration of local lending patterns has been a
major feature of research on building societies in
the United Kingdom (Boddy, 1976a; 1976b; 1980;
Duncan, 1976; 1977; Ford, 1975; Harrison and Stevens,
1980; Karn, 1976; Bassett and Short, 1980b;
Williams, 1976; 1978). In essence this work has
pointed to the variations in lending within cities
giving support to arguments about both redlining and
disinvestment/reinvestment processes. The selectivity of lending decisions in relation to particular
types of properties or households has been exposed,
leading to attempts in a number of studies to develop
the account further by interviewing branch managers.
This latter phase has not only given insights into
the opinions managers held but also has thrown some
light on the way mortgages are allocated (e.g.
Equal Opportunities Commission, 1978; Housing
Monitoring Team, forthcoming; Rigge and Young, 1981).
Of course, in seeking to solve questions related to
why allocations take the form they do, the complexity
of the process began to be uncovered and questions
raised about the extent of managerial discretion.
In some accounts that discretion is given a substantial role (e.g. Ford, 1975; Rigge and Young, 1981);
in others it is much reduced (Boddy, 1976a). As we
have sought to argue at the outset this is largely a
function of the framework adopted. Both discretion
and constraint are observable in local lending
situations.
     Data were obtained from seven societies operat-
ing in Dudley (some with more than one branch) and

245

Peter Williams

interviews were held with fourteen managers
(regional and branch).   Details of the patterns of
lending are given elsewhere (Housing Monitoring Team,
forthcoming; Doling and Williams, 1980).   Briefly,
data were collected for 2,285 mortgage allocations
made during the period 1975-7.   These represented
probably at least 10 per cent of all mortgages
granted in Dudley over the three years.   The major
weakness of the data set was that there was no information on refusals.   However, given that, and
problems regarding missing information, the data
were analysed with respect to variations within and
between societies.   A variety of procedures were
used, including regression, showing conclusively the
extent of variation in lending patterns.   Each
society's branch or branches were considered with
respect to the type, price and location of property
and the previous tenure, socio-economic group, age
and income of the borrower.
     On each variable substantial variation was
apparent.   More to the point a pattern emerged.
Thus one national society was seen to maintain a
relatively up-market profile of lending through both
its branches.   Relatively, it lent less on pre-
1919 property or to persons of lower incomes, and,
following in part from that, many of its borrowers
were previous owners (53 per cent and 60 per cent
respectively).   By contrast, another national
society operated in general at a lower point in the
market.   Only 29 per cent and 26 per cent of
lending by both its branches were to previous owners
and it lent heavily to first-time buyers and on pre-
1919 property.   The performance of other societies
also had a certain stability, i.e. an identifiable
lending profile.   The two regional societies in
the sample tended to operate quite conservatively
but less so than the one society which was seeking
a national base.   The small local society's performance was interesting.   It lent mainly to
previous owners though many of these had low incomes.
     The initial explanation for the results must
lie in the market position of each society.   The
most conservative national society had a well-
established position in the town.   It was assured
of a flow of funds and had a substantial clientele
who had already purchased houses with its assistance.
The least conservative national society had only
opened its branches recently and was still seeking
to develop its position.   The society pursuing
national ambitions also lent conservatively, reflecting perhaps its more tenuous local base and its

246

relatively unstable financial structure.   Finally
the local society, with a declining base in the face
of competition, lent to its own ageing clientele in
the locality of the office.   It performed an inevitably diminishing local role.

These different lending profiles were investigated further by a consideration of the size of
deposit required, the percentage advance made and
the income ratio applied.   This showed that the
treatment received by similar groups varied substantially between societies and to a lesser degree
between branches of the same society.   Thus first-
time buyers with an income of between £3000 and
£3999 per annum buying post-1960 property in the
price range £6000 to £12,000 received advances
ranging from 79 per cent to 94 per cent.   The same
magnitude of variation was apparent with regard to
the income ratio used for this category of buyers
(it ranged from twice to three times recorded
income).

The data analysis suggested that the performance
of societies varied substantially within the local
market and it was argued that this was, in part, a
consequence of their market position.   It also
indicated that, given that position, managers within
those branches exercised some discretion in undertaking their task of allocating mortgages.
Certainly, their control over what sort of clientele
approached them for assistance varied substantially.
In the case of new managers in new branches their
options were quite narrow - a problem confirmed in
interviews reported elsewhere (Housing Monitoring
Team, forthcoming).   However, although they may
not be able to control the flow of clientele, the
managers are able to vary the terms of individual
cases and apply different ratios, deposit requirements and grant different percentage advances.
The impact of this upon individuals is obviously
substantial.   Thus, though the opportunities for
discretion are largely removed by the requirements
of prudent financial management and the structure of
the financial market, and by their own society's
position in that market and the housing market,
managers still have room to manoeuvre and adjust
their lending decisions in the light of their own
understanding of the role they are to play.   To
assume a standard response is to assume equal knowledge, ignore cultural and class differentials, and
deny structural variations in the position of institutions and the varied impact of changing market
structures.

Conclusion

In this chapter, the focus has been upon questions related to the process surrounding the structuring of urban areas and in particular the issue of residential mortgage allocation. Set within a perspective which seeks to delimit the role of market processes and institutional structures on urban outcomes the chapter has explored the place of building societies and building society managers in local housing markets.

It was argued at the outset that the managerialist perspective, whilst weak in many respects, at least focused attention upon an array of institutions and agencies which have a role in shaping the form of cities. It also marked an important step in raising questions regarding constraint and conflict, and institutional rather than individual decision-making. As has been argued elsewhere (Saunders, 1980; Williams, 1982) the weaknesses of managerial or institutional urban research must not overshadow these advances. In the light of critiques, largely deriving from Marxist urban political economy, a restructuring of perspectives used in analysing institutional activity was called for. As outlined, perspectives drawn from the literature on the political economy of organisations can be helpful here in assisting the process of respecifying broader Marxist concepts. Following Lukes' statement (1974), 'social life can only properly be understood as a dialectic of power and structure, a web of possibilities for agents, whose nature is both active and structured, to make choices and pursue strategies within given limits which in consequence expand and contract over time', it would seem essential that urban geographic research maintains and develops its interest in institutional roles through both theoretical and empirical work - a view recently endorsed by the Social Science Research Council-sponsored seminar series 'Social Geography and the City' (Herbert, 1980).

The account presented here attempted to tackle theoretical and empirical questions. Following a reappraisal of the conceptual framework, building societies were selected for specific empirical examination. It was argued that insufficient attention had been given to the origins, development and present structure of the building society industry when seeking to explain mortgage allocations. A whole series of market situations were

identified which demonstrated the highly complex
and conflicting environment in which these institutions operate.   This position was replicated at the
local level in which branch offices function.   Here
patterns of allocation were considered and shown to
be highly variable even between similar types of
societies.   While part of this variation could be
explained by higher level constraints, e.g. the
society's position in the finance market, a role for
individual discretion was identified.   The direction in which this discretion was exercised relates
to the background and experience of the individuals
concerned as well as their institutional context.
The broad ideological statements made by managers
about supporting the growth of owner-occupation or
treating each case on its merits in reality masks a
whole series of different perceptions about needs
and responsibilities.   In constructing accounts of
institutional activity such dimensions should not
be ignored and their ideological dimensions remain
unexposed.   While this chapter could only direct
attention towards these issues their importance must
not go unrecognised.

     The implications of this assessment for both
theoretical structures and methodology are quite
substantial.   More attention must be directed
towards conceptual frameworks and the way these shape
and guide analytical work.   The tendency in
geography to undertake empirical research with an
implicit rather than explicit theoretical framework
can be seen to be unhelpful here.   Having said
that, the complexities of constructing consistent
and sufficiently illuminating theory cannot be
understated.   Paralleling the need for greater
theorisation is the necessity to develop methodologies appropriate to answering the kinds of questions
raised.   Theory and method interrelate and, in
adopting particular stances, particular methods are
implied.   A concern therefore with the exposure of
questions related to ideology and politics cannot
simply be satisfied by conventional survey research
which generates numerical estimates of particular
phenomena.   Interviews and participant observation,
when conducted within a specific framework, offer up
considerable possibilities as work in politics,
sociology and social history demonstrates.   Thus
the works of Saunders (1979), Paris and Blackaby
(1979) and Henderson et al. (forthcoming) come much
closer to being able to tackle questions about the
processes which structure local decisions because in
each case the authors seek to penetrate the

organisations concerned rather than simply to observe the outcomes of their actions.   As Saunders comments (1979, p. 348):

> When we consider the sorts of research techniques which are most likely to be available in studies of power relations, it soon becomes apparent that they are all characterized by the ability of powerful interests to keep the researcher at arms length.   Documentary sources, interviews, questionnaires, observation - all this will tend to produce a fairly formal picture even though the aim of such research is precisely to transcend the formal facade of local politics.

> The one technique which appears to offer the best chance for overcoming this difficulty is itself probably the most awkward to use in the analysis of political power, and this is participant observation.

The challenge for urban research in general and urban geography in particular is to develop its conceptual and methodological apparatus.   This will necessitate the final abandonment of any idea of an autonomous urban realm.   Mingione (1981) has argued recently that the perseverance of that view has ensured the continuation of functionalist methodologies and compelled the 'scholar to assume a set of independent variables whose influence on the object is finally explained as a mechanical and functional link'.   At the same time he stresses the dangers of developing typologies and ignoring historical dimensions.   It has only been possible to develop some of these issues within this chapter which should be seen, in part, as a contribution to the continuing process of debate and change within geography.   What is quite clear is that the institutional or organisational focus set within context and related to the 'dialectical process of capital accumulation' (Mingione, 1981, p. 70) is an entirely appropriate strand of geographic research.

References

Bassett, K. and Short, J.R. (1980a) Housing and Residential Structure: Alternative Approaches, Routledge and Kegan Paul, London
Bassett, K. and Short, J.R. (1980b) 'Patterns of

Building Society and Local Authority Mortgage
Lending in the 1970s', Environment and Planning
A, 12, 279-300

Bellman, Sir C. H. (1927) The Building Society
Movement, Methuen, London

Boddy, M. (1976a) 'Building Societies and Owner-
Occupation' in Housing and Class in Britain,
Political Economy of Housing Workshop, Confer-
ence of Socialist Economists, London, pp. 30-43

Boddy, M. (1976b) 'The Structure of Mortgage
Finance: Building Societies and the British
Social Formation', Transactions, Institute of
British Geographers, New Series, 1, 58-71

Boddy, M. (1980) The Building Societies, Macmillan,
London

Cleary, E.J. (1965) The Building Society Movement,
Elek, London

Clegg, S. and Dunkerley, D. (1980) Organization,
Class and Control, Routledge and Kegan Paul,
London

Committee to Review the Functioning of Financial
Institutions, (1980) Final Report, Her
Majesty's Stationery Office, London

Community Development Project (1976) Profits against
Houses, CDP Information and Intelligence Unit,
London

Davies, G. and Davies, M. (1979) 'Building Society
Branching Programmes will Continue for Several
Years', Building Societies Gazette, October

Doling, J. and Williams, P. (1980) 'Building
Societies and Local Lending Behaviour', mimeo,
Centre for Urban and Regional Studies, Univer-
sity of Birmingham

Duncan, S. (1976) 'Self-help: The Allocation of
Mortgages and the Formation of Housing Sub-
markets', Area, 8, 307-16

Duncan, S. (1977) 'Housing Disadvantage and Residential Mobility: Immigrants and Institutions in
a Northern Town', Working Paper 2, Urban and
Regional Studies, University of Sussex,
Brighton

Engels, F. (1970) The Housing Question, Progress
Publishers, Moscow.

Equal Opportunities Commission (1978) 'It's Not Your
Business, it's how the Society Works.' The
Experience of Married Applicants for Joint
Mortgages, EOC, London

Ford, J. (1975) 'The Role of the Building Society
Manager in the Urban Stratification System;
Autonomy versus Constraint', Urban Studies, 12,
295-302

Gauldie, E. (1974) Cruel Habitations; a History of
    Working Class Housing 1780-1918, Allen and
    Unwin, London
Gibson, K. (1981) 'Structural Change within the
    Capitalist Mode of Production: The Case of the
    Australian Economy', unpublished PhD thesis,
    Department of Geography, Clark University,
    Worcester, Massachusetts
Godsden, P.H. (1974) Self-help, Batsford, London
Gough, T.J. (1979) 'Building Society Managers and the
    Size Efficiency Relationship', Applied
    Economics, 11, 185-94
Gough, T.J. and Taylor, T.N. (1979) The Building
    Society Price Cartel, Hobart Paper No. 83,
    Institute of Economic Affairs, London
Gray, F. (1976) 'Non-explanation in Urban Geography',
    Area, 7, 228-35
Gregory, D. (1981) 'Human Agency and Human Geography',
    Transactions, Institute of British Geographers,
    New Series, 6, 1-18
Harrison, M. and Stevens, L. (1980) Basic Data on
    Local Authority Lending for Private House
    Purchase, and on the Local Authorities/
    Building Societies Support Scheme.   Housing
    Research Working Paper 3, Department of Social
    Policy and Administration, University of Leeds
Harvey, D. (1973) Social Justice and the City, Edward
    Arnold, London
Harvey, D. (1974) 'Class Monopoly Rent, Finance
    Capital and the Urban Revolution', Regional
    Studies, 8, 239-55
Harvey, D. (1975) 'The Political Economy of Urbanis-
    ation in Advanced Capitalist Societies: The
    Case of the United States', in Gappert, G. and
    Rose, H.M. (eds.), The Social Economy of Cities,
    Urban Affairs Annual Reviews, Vol. 9, Sage,
    Beverly Hills, California, pp. 119-63
Harvey, D. (1977) 'Government Policies, Financial
    Institutions and Neighbourhood Change in United
    States Cities' in Harloe, M. (ed.), Captive
    Cities, Heinemann, London, pp. 123-39
Harvey, D. (1978) 'The Urban Process under Capital-
    ism: A Framework for Analysis', International
    Journal of Urban and Regional Research, 2, 101-
    31
Herbert, D. (1980) 'Social Geography and the City',
    a report on an SSRC Seminar series, Department
    of Geography, University College Swansea
Henderson, J. et al. (forthcoming), 'Race and the
    Allocation of Council Housing', Final Report to
    the Department of the Environment, Centre for

Urban and Regional Studies, University of
Birmingham
Housing Monitoring Team (forthcoming) 'Building
Societies and the Local Housing Market', Centre
for Urban and Regional Studies, University of
Birmingham
Johnston, R.J. (1980) City and Society, Penguin,
Harmondsworth, Middlesex
Karn, V. (1976) 'Priorities for Local Authority
Lending: A Case Study of Birmingham', Research
Memorandum 52, Centre for Urban and Regional
Studies, University of Birmingham
Kemp, P. (1980) 'Housing Production and the Decline
of the Privately-rented Sector; Some Preliminary
Remarks', Working Paper 20, Urban and Regional
Studies, University of Sussex, Brighton
Lambert, C. (1976) 'Building Societies, Surveyors
and the Older Areas of Birmingham', Working
Paper No. 38, Centre for Urban and Regional
Studies, University of Birmingham
London, B., Bradley, D.S. & Hudson, J.R. (eds.)
(1980) 'The Revitalization of Inner-city
Neighborhoods', special issue of Urban Affairs
Quarterly, 15, 369-501
Lukes, S. (1974) Power, a Radical View, Macmillan,
London
Mingione, E. (1981) Social Conflict and the City,
Blackwell, Oxford
Pahl, R.E. (1970) Whose City? Longman, London
Pahl, R.E. (1979) 'A Comment', Area, 11, 88-90
Paris, C. and Blackaby, R. (1979) Not Much Improve-
ment, Heinemann, London
Pawley, M. (1978) Home Ownership, Architectural
Press, London
Registry of Friendly Societies (1979) Grays Building
Society: An Investigation under Section 110 of
the Building Societies Act 1962, Cmnd. 7557,
Her Majesty's Stationery Office, London
Rigge, M. and Young, M. (1981) Building Societies
and the Consumer, National Consumer Council,
London
Rose, H. (1978) 'The Housing Policy Review and the
Finance of Owner Occupation', Building
Societies Association, London
Royal Commission on Friendly and Benefit Building
Societies (1872) Second Report, Parliamentary
Papers, Her Majesty's Stationery Office, London
Saunders, P. (1979) Urban Politics, Hutchinson,
London
Saunders, P. (1980) 'Towards a Non-spatial Urban
Sociology', Working Paper 21, Urban and

Peter Williams

Regional Studies, University of Sussex,
Brighton

Smith, N. (1979) 'Gentrification and Capital:
Practice and Ideology in Society Hill',
Antipode, 11, No. 3, 24-35

Tomer, J. (1980) 'The Mounting Evidence on Mortgage
Redlining', Urban Affairs Quarterly, 15, 488-
501

Williams, L. (1981) Report on a speech by Leonard
Williams, Chairman of the Building Societies
Association, Building Society News, Building
Societies Association, London

Williams, P. (1976) 'The Role of Institutions in
the Inner London Housing Market: The Case of
Islington', Transactions, Institute of British
Geographers, New Series, 1, 72-82

Williams, P. (1978) 'Building Societies and the
Inner City', Transactions, Institute of British
Geographers, New Series, 3, 23-34

Williams, P. (1982) 'Restructuring Urban Managerialism: Towards a Political Economy of Urban
Allocation', Environment and Planning A, 14,
95-105

Woolwich Review (1981) 'Planning Appeals', Woolwich
Equitable Building Society, London

Chapter 9

THE LOCAL STATE AND THE JUDICIARY: INSTITUTIONS IN AMERICAN SUBURBIA

R.J. Johnston

The crucial role of the state in the regulation of economic and social life, and thus in the creation, maintenance and alteration of spatial structures, has only recently been appreciated by geographers (Cox, 1980; Johnston, 1979; 1981a; in press; Burnett and Taylor, 1981). Empirical realisation has been associated with theoretical speculation concerning the nature of the state, as geographers seek to understand the why and where of state activity (Dear and Clark, 1978; 1981; Dear, 1981). Incorporation of the political element is necessary, it is argued, for a full appreciation of the forces structuring the geography of late capitalist societies.[1]

Theoretical writings on the state advance a variety of interpretations of its nature and functions (see Greenberg, 1979; Saunders, 1979; Dunleavy, 1980; Johnston, 1982a). The structuralist interpretation, which is accepted here, is that the state exists in a capitalist society to maintain the necessary conditions for capital accumulation and the related social formation. To do this, according to O'Connor (1973; see also Saunders, 1979), the state is involved in three types of activity and expenditure: (1) social investment in projects that will enhance labour productivity and thereby sustain private capital accumulation; (2) social consumption, which reduces the cost to capital of the reproduction of the labour force by either providing or subsidising the material and cultural conditions of existence; and (3) social expenses on the maintenance of harmony.

Capitalist society is class-based. Within the working class there is considerable competition and conflict over access to the better-rewarded occupations and to the pleasanter living

R.J. Johnston

environments.   Such access is strongly influenced
by state actions in, for example, the provision of
education and the creation of residential areas.
Thus the conflict within capitalist society
necessarily involves the state in its social investment and social consumption functions.   Regulation
of that conflict involves the state in its maintenance of harmony function.

The Local State

The state is not a single unit.   In most countries
it is subdivided, both sectorally and territorially.
The sectoral subdivisions allocate supervision of
separate aspects of social and economic affairs to
various government departments and agencies: the
territorial subdivisions involve various functions
being provided for separately defined jurisdictions
within the geographical whole.   Both sets are subject to overall central control, and they interact
with each other, some sectoral subdivisions having
their own sets of territorial subdivisions.
     For many people, it is the territorial sub-
divisions which are the immediate representations of
the state, those with which they are most likely to
come into contact during their daily lives and
which provide the public sector goods and services
they most frequently use.   Indeed, in many
countries the territorial subdivisions are quanti-
tatively the most important elements of the state.
In the United States in 1977, for example, the
expenditure of the Federal government was $382
billion (throughout this chapter, one billion is
one thousand million) whereas that of state and
local governments was $321 billion; the Federal
civilian employment was 2.86 million (there were a
further 2 million military personnel on active duty)
and the state and local governments together
employed some 12.8 million persons (82 per cent of
the total non-military employment by government).
Payments to the Federal government by individuals
were mainly in the form of income taxes ($157.6
billion) and social insurance contributions ($108.7
billion): state and local governments raised $175.9
billion in taxes, mainly on property ($62.5 billion)
and on sales ($60.6 billion).   Finally, although
there is only a single Federal government, there
were 50 state governments and 79,862 local governments.   Many of the latter overlap, so that an
individual American may be a resident of, and paying

taxes to, several separate territorial authorities.

The size of the local government sector and its importance to the daily lives of most Americans has generated considerable theoretical as well as empirical interest in what has become known as the 'local state' (Dear, 1981). Within this work there has been considerable debate concerning the autonomy of the local state. Are American local governments independent units, within the overall sovereignty of the United States, or is their freedom to act considerably constrained? Dear and Clark (1981) have concluded from their study of Massachusetts that the latter interpretation is correct; nevertheless they agree with Dunleavy (1980, p. 134) that local politics should be studied, whilst accepting the possibility that

> local politics is fundamentally epiphenomenal or surface activity, reflecting the playing out of much broader social forces in particular spatial and institutional contexts, but not itself encapsulating any effective determining influences on urban policy-making.

Both the empirical investigation and the theoretical articulation of the local state are still in their infancy. Much remains to be done in the matter of uncovering how the local state operates, how much independence it has, and how decisions are made within it. Theoretical advances will be made through further investigation of the state and its links with the capitalist political economy (Taylor, 1980; 1981). Empirically, investigations are needed into the particular circumstances of different times and places. The latter is the aim here. The focus is on one type of local state - the municipality - in one territorial segment of the USA - suburbia - with regard to one activity - the control of land use. Because some of the policies of land-use control are enacted to control the nature of local schools, however, some reference will be made to another element within the local state - the school district.

As already indicated, the local state exists in the USA as a creation of a higher level institution - one of the fifty states of the Federation. Its autonomy is prescribed at that higher level, where it can be either altered or removed - either by legislative action or by constitutional amendment.

The Judiciary

Few actions of the government of a local state will
meet with universal approval from either the
residents of its territory or the residents of
other territories. Many who disagree with a
decision will accept it. Some, however, may
initiate a challenge, either to get the particular
decision reversed or to get the constraints on
local authority altered so that such a decision
cannot be made again, there or elsewhere.
Such challenges to local state activity may
take one of three forms:

1. local political activity aimed at reversing
   a particular decision - this may involve a
   range of actions which Cox and McCarthy
   (1980; in press) have grouped together as
   neighbourhood politics; in the thirty-nine
   states where the popular initiative is
   allowed at the local level (Ranney, 1978),
   a referendum may be demanded;
2. more general political activity to get the
   constraints to local autonomy altered -
   this may involve seeking legislative action
   in the state capitol including, in the
   fourteen states where it is relevant
   (Ranney, 1978), petitioning for a statewide
   referendum on the issue; and
3. a challenge to the legality of a particular
   decision in the courts - if successful,
   such a challenge may create precedents for
   the prevention of similar decisions elsewhere. All of these are widely used in
   the United States with respect to the
   local state. Only the third type is discussed here.

Each state has its own courts, which interpret
the state constitution and to which grievances can
be taken and cases brought against local governments.
Thus the state courts are forced, when required by
litigants, to intervene in particular conflicts;
the first task for the courts is to ensure that the
actions brought before them are proper and that
they have jurisdiction over such conflicts. In
addition, cases involving the Federal constitution
may be appealed to the Federal courts if one or more
of the litigants is not satisfied with the decisions
of the state courts; certain cases can be taken

directly to the Federal courts.   The highest
Federal court is the United States Supreme Court,
which is the final court of appeal.   The Supreme
Court hears only a small proportion of the cases
appealed to it (Hodder-Williams, 1980), but many of
its decisions set precedents for the interpretation
of the Federal constitution and these constrain
judicial decision-making in both the lower Federal
and the state courts, as well as in the Supreme
Court itself.

The judiciary is a part of the state, but in
the United States, as in many other countries, one
of its principal characteristics is its independence
of the other arms of the state (the executive and
the legislature) and of various sections within
society.   This independence is guaranteed (at the
level of appearance - Clark, 1981) by salary levels
and by the security of tenure enjoyed by the justices.   (Federal judges 'shall hold their offices
during good behavior' according to the US
Constitution.)

The need for a system of courts to interpret
the (Federal and state) constitutions reflects both
the lack of clarity which many people discern in
such documents, especially given the changing nature
of society and economy and the relative permanence
of constitutions.   How judges interpret constitutions (and laws, which must be constitutional)
depends in part upon their personal attitudes and
value-systems.   The flexibility of interpretation
of such written documents means that judges frequently differ in their opinions, so that which
judges are called upon to rule in a particular case
can be a crucial influence on the outcome of that
litigation.   (Judges are constrained by legal
precedent - prior rulings on similar or related
issues - but again many precedents are so written
that they are open to more than one subsequent
interpretation.)   Most judges tend to interpret
constitutions, laws, and precedents in line with
their own ideologies, however.

Judicial appointees to the Federal courts are
nominated by the President to the Senate for
approval so that Presidents almost invariably nom-
inate judges, especially Supreme Court judges, who
share their own ideologies (Simon, 1973).   They
are placed in the courts to interpret the law in
the way that the nominating President wishes.   However, because of their security of tenure, the
judges on the Supreme Court at any one time may not
reflect the views of the current incumbent of the

259

White House (Jackson, 1941; Swindler, 1970). By
their nominations, when the opportunities arise,
Presidents are able to influence the ideological
composition of the courts, but they put no direct
constraints on judicial decision-making.

The Local State and the Judiciary

Two types of institution have been isolated for
study here - the local state and the judiciary.
Neither is independent of the encompassing institution of the state (which in its turn is not independent of the contemporary political economy: Jessop,
1977). But each has some freedom to act within
specified legislative and constitutional constraints,
and that freedom and how it is used is the subject
of the present essay. In addition, and significantly for the discussion here, the two institutions interact, in that one - the judiciary - may
be called upon to resolve conflicts involving, as at
least one party, the other - the local state.
    The range of subjects that could be considered
relevant to the interaction between local state and
judiciary is large. The focus here is on one
topic only and on only one element of the judiciary:
the regulation of land use within its territory by
the governing body of the local state, with particular reference to suburban municipalities. The
interaction involves the decisions of the US
Supreme Court that are either directly related to
land-use regulation or affect that activity through
legal precedent. The discussion does not provide
a full outline of the nature of suburbia in
American metropolitan areas (see Muller, 1980;
Walker, 1978); it illustrates how institutions can
be used in the manipulation of suburban environments.

The Local State in the USA

The nature of local government in the United States
is determined by the states, all of which have
clauses relating to the devolution of functions to
territorially defined jurisdictions as parts of
their constitutions. (Local, as opposed to state,
government is not mentioned in the United States
Constitution.) Most states are exhaustively
divided into counties. Within these, municipalities can be created as independent territorial

units for certain functions.   There are variations
between states in the procedures for incorporation,
but in most the process involves residents of the
relevant territory petitioning the state legislature, which in turn organises a referendum of all
inhabitants of the defined area (Hallman, 1977).

In addition to the counties and municipalities,
there are two other forms of local government
common in the states.   The first are the school
districts, which are ad hoc territorial units that
provide primary and secondary education only.   In
some states, education is provided by county governments and by some municipal governments - especially
those of the largest cities - but most of the state
territory is covered by a mosaic of school districts
whose boundaries frequently do not coincide with
those of municipalities.   Finally, there are the
special districts, ad hoc territorial authorities
created for a wide range of single functions.   The
reasons for their creation vary considerably
(Stetzer, 1975).

In total, there were 79,862 local government
units in the United States according to the 1977
Census of Governments, of which 67,780 had the
power to raise revenue through property taxes.
The total was divided into: counties, 3,042;
municipalities, 18,862; school districts, 15,174;
special districts, 25,962; and townships, 16,822.
(Townships are sub-county units in twenty states of
the Northeast and Midwest; their powers are few.)
Almost one-third of these units were in the Standard
Metropolitan Statistical Areas, as defined in 1977.
For the various types the percentages in metropolitan areas were: counties, 19.5; municipalities,
34.2; school districts, 33.1; special districts,
36.9; and townships, 24.0.   Many units are small.
The modal metropolitan school district in 1977, for
example, operated between three and nine schools,
for between 3,000 and 5,999 pupils; the modal intra-
metropolitan municipality had a population of less
than 1,000.(For fuller details on the composition
of the intra-metropolitan governments, see
Johnston, 1981b.)

Of these various local government units, the
main focus of the present essay is the suburban
municipality, of which there were just over 6,000
in the 272 Standard Metropolitan Statistical Areas
in 1977.   The rationale for their existence, their
powers, and the challenges to their operations is
discussed below.   However, in most metropolitan
areas the suburbs have a mosaic of school districts

in addition to a set of incorporated municipalities
plus unincorporated county areas.   This necessitates some discussion of the educational as well as
the municipal balkanisation of American suburbia.

The Operation of the Local State

A capitalist society is a class society, and is
characterised by continuous inter-class and intra-
class conflict over the allocation of rewards to
different groups within the division of labour.
Much of this conflict takes place over the built
environment (Harvey, 1978); it results in territorial distancing between classes, for social and economic reasons (Johnston, 1980; 1982b).   Much of the
rationale for distancing, for people of similar
social and economic status congregating together
and segregating themselves from other groups, concerns members of particular sections of society
seeking to protect their positions.   Most individuals and households which have attained relatively
high status aim to protect this and to try and
ensure it for their children, whereas others may
seek to better themselves and their offsprings'
prospects.   To achieve these aims, certain residential milieux are considered more desirable than
others: they provide acceptable environments for
child socialisation; they contain schools which are
more likely to produce successful graduates for
the labour market; and they protect the residents'
investments in property values.
     In the conflict over the built environment, it
is the aim of the already successful to insulate
themselves from those considered their social and
economic inferiors whilst many of the latter seek
to improve their situation - and the potential
situation of their children.   Thus classes are in
conflict over residential areas, in a conflict
which transcends solely economic issues.
     Associated with this class conflict in many
societies is ethnic conflict.   The urbanisation
process which accompanied industrialisation required
the congregation of large numbers of people in
large settlements to provide the needed labour
force for capitalist advancement.   In many
societies, such urbanisation involved bringing together groups from different cultural backgrounds.
Many of these were rapidly assimilated into the new
urban society and lost most traces of their former
identity.   In most cases, however, at least one

group was constrained to the lowest positions in the division of labour, thereby forming both a caste-like group who were allowed to perform only the most menial tasks, and were rewarded poorly for it, and an industrial reserve army which suffered earliest and most in any economic recession.   Thus the process of industrial-urbanisation was frequently associated with racialism.   Groups with separate racial identities were obvious targets for such discriminatory policies simply because of their different appearance.

In the United States, this group during the nineteenth and twentieth centuries has been the descendants of the West Africans imported in the years prior to, and just after, independence to provide slave labour on the southern plantations. During the twentieth century, large numbers of blacks have moved from the rural south to the big cities in the search for work.   This search has been hampered by restrictive practices.   Even more discriminatory, however, have been the restrictive practices operated in the housing market.   The general belief of the white population has been that blacks are inferior.   Thus social contact with them has been considered undesirable, and in particular white families have sought to ensure that their children do not attend the same schools as blacks.   Because American public school systems are organised on a neighbourhood basis, this means that spatial separation is desired by the whites on ethnic as well as class grounds.   Thus in addition to the class cleavage within American society there is also a racial cleavage which is reflected in conflicts over the built environment.

The fragmentation of local government in suburbia is part of the American solution to this conflict over the built environment and provides a mechanism for the manipulation of space by powerful groups within society.   The creation of separate municipalities allows the residents of each incorporated area control over land use, and thus (indirectly though fairly closely) over the social and economic character of the residents.   The existence of separate school districts for these areas allows the control of the residential characteristics of suburbia via zoning to influence the socio-economic and racial composition of school populations.

Suburbanisation in the United States has not always been associated with the incorporation of separate municipalities but, as Walker (1978) makes

clear, a major aim has always been to allow the
better-off in urban society to congregate in pleasant physical and social environments, from which
lower class and certain ethnic groups are excluded.
The forces driving the wealthy to the edge of the
built-up area included the physical negative
externalities of noise, congestion, pollution and
health hazards of the densely occupied, increasingly
industrialised areas of the inner city and the
social negative externalities created by the large-
scale immigration of European and black labour to
the jerry-built tenement blocks close to the city
centre.   Movement to the suburbs was made possible
by developments in transport technology, allowing
the desired distancing to occur.

Initially, the suburbanites were quite happy
for their newly settled areas to be annexed by the
cities from whose centres they were fleeing; a
major consideration in a majority of cases was that
without annexation the suburbanites would have been
unable to obtain gas, electricity, water and other
public utilities (see Teaford, 1979, Chapter 3).
In some cases, annexation followed incorporation.

If the late nineteenth century was characterised by the annexation of suburbs into central
cities - especially those of the country's manufacturing belt - the twentieth century has become an
era of opposition to annexation.   Several reasons
account for this change, among them the increased
ability of small municipalities to provide utilities
and services for themselves, to obtain them from
the encompassing county government, to buy them from
either neighbours or private contractors, or to
combine with neighbours in the creation of special
districts designed to provide a single service only.
These were necessary but not sufficient conditions
for opposition to annexation, however.   As Gordon
(1977) has made clear, increasingly industrialists,
developers of residential areas, and residents became aware of the benefits to be derived from
separate incorporation and municipal independence
from the central city of the burgeoning metropolitan
area.   These include:

1. Because most local governments base their
    revenue-raising on property taxes,
    residents of separate suburban municipali-
    ties are not required to contribute to the
    budgets of either the central city govern-
    ments, or other parts of suburbia.   As
    the costs of servicing a densely occupied

commercial and industrial city are generally much greater than those for a low-density, high-income residential area (in providing police and fire protection, in the removal of garbage, in regulating industry, for example, and - increasingly in recent years - providing welfare benefits for the poor) municipal separation means opting out of fiscal responsibilities. Indeed, since more suburban residents use central city facilities (roads, public parks, public buildings, etc.) than vice versa, they are able to free-ride on the property tax contributions of those remaining in the central city (many of whom, because of the operations of the property and housing markets, are unable to afford a move to the suburbs).

2. When the suburban areas were still part of rural America, school districts were established to provide educational facilities for the farming communities. Annexation of these areas almost invariably involved annexation of the school districts into the city's school system. Separate municipal incorporations, however, meant that the school districts were retained. Thus suburban residents were not called on to contribute to an education system for the entire metropolitan area; they paid only for their own local schools (and maintained close control over their operations). Because most suburban residents were more affluent than those in the central city, this meant that the former could avoid contributing to the costs of educating children of the relatively poor. This both reduced their tax burden (relative to the situation if annexation had occurred) and meant that central city school systems were relatively underfinanced; the chances of the children of the rich in the 'educational rat race' were thus enhanced.

3. Government of municipalities is by elected bodies, and in many of the large central cities competition between vested interest groups for control of the city government (often via the main political parties) frequently meant that the affluent middle class was defeated by a combination of

immigrant voters and big business. In
small suburban municipalities it was
possible to avoid the operations of
political machines, such as that of New
York's Tammany Hall, and elect a government more 'in tune' with the wishes of
the relatively affluent. In the late
nineteenth and early twentieth centuries,
this desire to avoid the politics of big-
city government produced a 'reform' movement which aimed to replace elected
councils, whose members represented wards
and canvassed support on a partisan basis,
by more efficient bodies. The most
common of these is the 'council-manager'
municipal government in which a small
council, elected at-large and on a non-
partisan ballot in most cases, oversees
the work of a city manager, who in turn
supervises the work of a series of departmental managers. The reform movement was
especially successful in the South and
West, among medium-sized municipalities
whose residents had above-average incomes
and educational levels.

4. Finally, the home rule granted to separately incorporated municipalities has given
them control over the use of land within
their jurisdictions, through zoning.
Some of the early incorporations were
undertaken for this purpose (Bigger and
Kitchin, 1952). Most zoning plans,
especially for the smaller, more affluent
suburban municipalities, are designed to
create a particular socio-economic milieu
(see below).

Not all of these advantages remain, for state
and Federal legislation has to some extent eroded
the home rule independence of suburban municipalities. The increasing Federal contribution to
education costs, for example, means that suburban
residents must pay part of the costs of central city
schools via Federal income taxes, and in several
states legislative and/or court decisions taken
under the state constitution require equalisation
of fiscal resources between school districts, so
that suburban residents must contribute to the costs
of education elsewhere through their state taxes
(Lawyers' Committee for Civil Rights under Law,
1980). In addition, local school boards must

conform to state guidelines on a variety of issues.
Control over land use has been retained, however, so
that residents of suburban municipalities can still
influence who lives within their jurisdictions and
this in turn determines who attends the local
schools provided by the independent suburban school
districts.

Managing the Suburban Milieux

A common analytical model of the state, politics
and democracy in capitalist countries presents a
pluralist view, in which various elite groups compete - usually through political parties - to
control the relevant government and to pursue
certain policies, the fruits of which will benefit
their supporters most.   No one group has exclusive
control over the government, so that none is always
the loser (for reviews see Saunders, 1979; Dunleavy,
1980).   This is not the case in American suburbia,
however, where the possibility of separate incorporation allows groups to insulate themselves from
pluralist conflict and establish local hegemony.
Many municipalities, and especially the smaller
ones, are controlled by a homogeneous group within
which there may be differences over means and ends
but these are relatively minor.   The balkanised
suburbs comprise a series of independent islands in
each of which there is considerable consensus over
goals.
     One of the major goals is the preservation of
the socio-economic environment; undesirables must
not be allowed in who would devalue the social
milieu, lead to a depreciation in property values,
and make large demands on the municipal budget.
To keep these people out (basically the relatively
poor, the blacks and other ethnic groups), many
municipalities have adopted policies of exclusionary zoning.   These involve either restricting the
volume of development altogether (new development
makes demands on the municipal budget, notably for
the provision of roads and utilities) or indirectly
restricting it to certain groups, whose contribution to the budget will be large.   As Danielson
(1976) makes clear in his detailed analysis of such
policies, this exclusionary zoning-sometimes termed
snob zoning - uses mechanisms such as large minimum
lot sizes, minimum lot dimensions, maximum areas
of the lot to be covered by buildings, expensive
building codes, and no provision for either

apartments or mobile homes.   Thus (p.73)

> The spread of land-use controls has ...
> increased the ability of local governments to
> influence informally the plans, prices and
> clientele of developers.   Just who is to be
> permitted to live in a community cannot be made
> explicit in a zoning ordinance or subdivision
> regulation.   But it can often be determined
> in negotiations between the local government
> and the developer.

Not all suburban municipalities operate 'snob
zoning' and aim to exclude all but very expensive
residential developments.   In many, it is the
middle income groups who are being attracted to live
there (although the poor are still being - implicitly -- excluded).   To help these people afford a
piece of suburbia, commercial and industrial zones
are created, to provide jobs and services and also,
importantly, to contribute to the property tax base.
Thus many suburbs are competing - among themselves
and with the central cities - for commercial and
industrial developments to guarantee fiscal
viability for their residents.   At the same time,
they plan these developments to minimise any negative externalities for residential areas, and their
zoning is designed to exclude those whose presence
might reduce the municipalities' attractions and
devalue their (actual and potential) tax bases.

Challenging Local State Managerial Policies in the Supreme Courts

Of the means of challenging the managerial policies
of the local state listed above, only the use of
the courts is considered here.   The focus is on
challenges which either directly or indirectly
relate to land-use zoning powers and on those which
have been appealed through the system to the Supreme
Court.   The number of relevant cases to be highlighted is small, but their importance with regard
to local state autonomy is considerable.
    Zoning decisions and other local state actions
relating to land uses within their territory can be
challenged on a variety of constitutional grounds.
The grounds most commonly used are given in the
Fourteenth Amendment, part of Section 1 of which
reads -

No State shall make or enforce any law which
shall abridge the privileges or immunities
of citizens of the United States; nor shall any
State deprive any person of life, liberty, or
property, without due process of law; nor deny
to any person within its jurisdiction the equal
protection of the laws.

The key phrases in this are 'due process' and 'equal
protection'.

This clause - generally known as the 'equal
protection' clause - was introduced after the Civil
War to define citizenship of the United States and
its rights and to ensure that the newly enfranchised
blacks were not denied their civil rights; the
amendment is entitled 'Citizenship, Representation,
and Payment of Public Debt'.   In the century or so
since its ratification (in 1868), however, it has
been used as the basis for a wide variety of causes
and has become the foundation for 'The growth of
intellectual and political support for an ideology
of laissez-faire, with its mixture of normative and
empirical arguments.'   (Hodder-Williams, 1980, pp.
135-6)

The debate over whether the framers of the
Amendment meant it to be widely interpreted has
been broad-ranging and the differences are deep-
rooted, between those who believe that the Constitution should be regularly re-examined in the light of
the changing social and economic environment and
those who believe that judges should stick with a
narrow interpretation stating what the intention of
the authors of the amendment was (Berger, 1977).
Which of these positions has been taken by the
judges at any one time has depended very largely on
the composition of the Court.   Judges have their
own ideologies.   Most of the cases relevant to
local state autonomy over zoning and land-use
matters have been heard by Courts dominated by
those who interpret the Constitution, and especially
the Fourteenth Amendment, relatively narrowly.

The Green Light for Zoning

Zoning was introduced to American urban areas in the
second decade of the twentieth century, being
pioneered in 1916 by the government of New York City
as a means of separating incompatible land uses in
densely occupied districts and of controlling industrial developments.   (For histories of zoning, see

Tull, 1969; Perin, 1977.) During the following
decade it was adopted by a large number of municipalities without its legality being tested. The
challenge which led to a ruling on its constitutional
validity was determined by the Supreme Court in
1926.

The case which came to the Supreme Court -
Village of Euclid v Ambler Realty Company 272 US 365,
71 L Ed 303 (1926)[2] - referred to the municipality
of Euclid, a suburb on the eastern boundary of the
city of Cleveland, Ohio. Ambler Realty Company
owned a block of land in Euclid which it intended to
develop for industrial uses: its assessed value was
$10,000 per acre. The municipality's zoning plan
allocated it for residential use, however, and
Ambler estimated that this reduced its value to only
$2,000 per acre. The suit was brought that this
decision denied the company its property rights
without due process and thus denied it the equal
protection of the law.

In its 6-3 majority decision in favour of the
Village of Euclid, the Court noted that the board of
zoning appeals - to which Ambler Realty Company had
initially appealed - was authorised to interpret
the zoning ordinance 'so that the public welfare
may be served and substantial justice done' (p. 309)
and it ruled that

> We are not prepared to say that the end in
> view was not sufficient to justify the general
> rule of the ordinance ... It cannot be said
> that the ordinance ... passes the bounds of
> reason and assumes the character of a merely
> arbitrary fiat (p. 311).

Thus although Ambler Realty Company had been denied
the full value of its property by a state decision,
this was not an arbitrary act of the municipality
but was undertaken to advance the public good. In
such a case, due process and equal protection notwithstanding, the rights of the individual could be
overridden. This ability to override individual
rights to advance general welfare is known as the
police power of the states, defined as 'the
authority of the state to govern its citizens, its
land and its resources, and to restrict individual
freedom to protect or promote the public good'
(Witt, 1979, p. 358).

The Rare Statements on Zoning

Having determined in 1926 that zoning land-use patterns was a valid use of the police power by the
state - which it delegated to local governments -
the Supreme Court from then on has played
relatively little part in adjudicating on issues
relating to the detail of zoning.   (In a later
decision - Nectow v Cambridge 277 US 183, 72 L Ed
842 (1928) - the Court ruled that any zoning
decision must clearly be shown to be in the general
welfare, otherwise due process was being violated.)
Because zoning is a state power, its operation only
comes under the Supreme Court's jurisdiction when
issues relating to the Federal Constitution, other
than those covered by the Euclid decision, are
raised.   American case law is built upon precedent,
so that once a definitive decision has been handed
down, further hearings on the same issue by the
Supreme Court are unnecessary.   Local courts may
have to handle a great volume of detailed cases
relating to particular zoning decisions (indeed,
zoning issues are among the most common causes of
litigation: Dolbeare, 1969), but the fundamental
validity of the process has been determined.
     Very few cases relating to zoning have been
heard by the Supreme Court since 1926, but it has
been involved in three types of case: those dealing
with racial discrimination in the access to housing,
which clearly fall under the Fourteenth Amendment
and set precedents for zoning policy; those dealing
with the criteria for zoning; and those dealing with
challenges to zoning.

Access to Housing and Residential Areas

On the issue of racial discrimination, the Court has
always ruled that zoning cannot be used as a direct
means of segregating blacks from whites (i.e. racial
issues cannot be a part of the zoning ordinances).
Prior to the Euclid decision, in 1917 the Court
ruled - in Buchanan v Warley 245 US 60, 62 L Ed 149
(1917) - against a 1914 ordinance by the city of
Louisville, Kentucky, section 1 of which made it
'unlawful for any colored person to move into and
occupy as a residence ... any house upon a block
upon which a greater number of houses are occupied
... by white people than ... by colored people'
(p. 159).   Section 2 similarly made it unlawful for

whites to move into dominantly black areas.
According to Mr Justice Day, speaking for the
Court,

> It is said such legislation tends to promote
> the public peace by preventing racial conflicts;
> that it tends to maintain racial purity; that
> it prevents the deterioration of property owned
> and occupied by white people, which deteriora-
> tion, it is contended, is sure to follow the
> occupancy of adjacent premises by people of
> color (p. 160).

But the case of Buchanan v Warley was not concerned with the equal protection clause; instead,
it related to the due process clause, having been
brought by a white who claimed arbitrary treatment
because he was unable to dispose of his property
(to a black).   As Mr Justice Day put it

> We think this attempt to prevent the alienation
> of the property in question to a person of
> color was not a legitimate exercise of the
> police powers of the state and is in direct
> violation of the fundamental law enacted in
> the 14th Amendment ... preventing state inter-
> ference with property rights except by due
> process of law (p. 164).

Thus racial zoning was invalidated without reference
to racial discrimination.   As Williams (1950)
points out, several other cities tried to avoid the
precedent set by this case but have been unable to
do so when challenged in the courts.
The inability of zoning bodies to use racial
classifications in their ordinances was made
explicit by the precedent set in a case referring to
Akron, Ohio - Hunter v Erickson 393 US 385, 21 L Ed
2d 616 (1969).   The Akron City Council had adopted
an ordinance, after a referendum, which stated that

> Any ordinance enacted by the Council of the
> City of Akron which regulates the use, sale,
> advertisement, transfer ... of any real
> property of any kind or of any interest therein
> on the basis of race, color, religion, national
> origin or ancestry must first be approved by a
> majority of the electors voting on the question
> (p. 620).

In 1964, prior to the passing of this ordinance,

Akron had passed a 'fair housing ordinance' prohibiting discrimination in housing but the later
ordinance, which was made retrospective to override
the fair housing one, denied a black (Mrs. Hunter)
the benefit of this without a referendum. She
claimed discriminatory treatment under the Fourteenth
Amendment, and the Supreme Court upheld her appeal
because the new ordinance contained an 'explicitly
racial classification' (p. 621).

Criteria for Zoning

On the issue of the criteria for zoning, the Supreme
Court has indicated that the local state has wide
discretion in interpreting the public good.   This
was indicated in its Euclid decision in a phrase
referring to 'depriving children of the privilege of
quiet and open spaces for play ' (p. 313) and was
further stated in a case relating to the operations
of the District of Columbia Redevelopment Land
Agency.   In Berman v Parker - 348 US 26, 99 L Ed
27 (1954) - it was claimed that the Federal Legislature was violating the Constitution (in this case
under the Fifth Amendment which states that 'no
person shall be ... deprived of life, liberty, or
property, without due process of law; nor shall
private property be taken for public use, without
just compensation') in passing an Act which allowed
the Agency to acquire a commercial property as part
of its residential redevelopment programme.   In
ruling in favour of the Agency the unanimous Court
stated that 'It is within the power of the Legislature to determine that the community should be
beautiful as well as healthy, spacious as well as
clean, well-balanced as well as carefully patrolled.'
(p.33)
        A further case dealing with this issue referred
to an attempt by a small municipality to prevent
students from an adjacent university campus - the
State University of New York at Stony Brook, on Long
Island - from renting houses there.   The residents
of the village of Belle Terre (an area of less than
one square mile with a population of about 700) did
this by introducing a zoning ordinance intended to
restrict occupancy of dwellings to nuclear families
only.   It stated that a family comprises

> One or more persons related by blood, adoption
> or marriage, living and cooking together as a
> single household unit, exclusive of household

273

servants.   A number of persons but not
exceeding two living and cooking together as a
single housekeeping unit though not related by
blood, adoption or marriage shall be deemed to
constitute a family.

This ordinance clearly outlawed any attempt by
groups of students to rent a home in Belle Terre and
in Village of Belle Terre v Boraas - 416 US 1, 39 L
Ed 2d 797 (1974) - a case was brought charging that
the ordinance inhibited the students' freedom of
the right to travel, that social homogeneity in a
neighbourhood was not a valid concern of government
and that the fact of marriage was irrelevant to
neighbours.   The Supreme Court found in favour of
the village, however, and in a decision echoing and
developing on the phrase quoted from the Euclid case
above, Mr Justice Douglas (for the majority of
seven judges) wrote

> The regimes of boarding houses, fraternity
> houses, and the like present urban problems.
> More people occupy a given space; more cars
> rather continuously pass by; more cars are
> parked; noise travels with crowds ... A quiet
> place where yards are wide, people few and
> motor vehicles restricted are legitimate guide-
> lines in a land-use project addressed to family
> needs ... It is ample to lay out zones where
> family values, youth values and the blessings
> of quiet seclusion and clean air make the area
> a sanctuary for people (p. 804).

In dissent, Mr Justice Marshall argued that the
ordinance did violate the students' rights guaranteed by the Fourteenth Amendment:

> Zoning officials properly concern themselves
> with the use of land ... But zoning authorities
> cannot validly consider who these people are,
> what they believe, or how they choose to live,
> whether they are Negro or white, Catholic or
> Jew, Republican or Democrat, married or un-
> married (pp. 807-8).

In their opinions on this case, however, seven other
judges disagreed - except over the issue of whether
residents are Negro or white - with Justice Marshall
that the right to establish a home where one wishes
is one of the fundamental rights guaranteed by the
Amendment.   (See, however, the related case of

Moore v City of East Cleveland - 431 US 494, 52 L Ed 2d 531 (1977) - in which an ordinance, overruled by the Court in a 5-4 decision, had the effect of denying the right of a woman to live in the same household as her sons and grandsons: Witt, 1979, p. 359). There is a great variety of local ordinances dealing with the details of land use - in 1976, for example, Palo Alto city council introduced one prohibiting residents from hanging out washing on clothes lines (The Times, 18 March 1981) - but only very rarely are they considered of such significance that cases contesting them must be heard by the Supreme Court.

One other case that did reach the Court concerned the refusal of zoning officials in the Illinois village of Arlington Heights to redesignate the allowable uses for a piece of land. A developer wished to build relatively high density, low- and moderate-income housing in the village, for both races. The village zoning board of appeals refused to allow such development, and in the case of Village of Arlington Heights v Metropolitan Housing Development Corporation - 429 US 252, 50 L Ed 2d 450 (1977) - it was claimed that the rights of black people under the equal protection clause had been violated. The Supreme Court disagreed, in a 5-3 ruling, because it found no evidence that the board's decision had been racially motivated. Direct proof of discriminatory intent was needed and this was not provided since all that the board had done had been to continue to apply, in a consistent manner, the zoning policy it had always followed, and no separate procedures had been adopted for this case. The Supreme Court did, however, refer the case back to the Seventh Circuit to see whether the village had been in violation of the Fair Housing Act. This latter court did identify a violation, because discriminatory effects did not require intent under the Act - Metropolitan Housing Development Corporation v Village of Arlington Heights 558 F2d 1283 (1977): 'We reaffirm our earlier holding that the Village's refusal to rezone has a discriminatory effect' (p. 1288) - thus indicating how legislative actions change the environment within which the courts operate.

Democracy and Exclusion

The Arlington Heights case concerned a challenge to the actions of a zoning board, which was overruled because no evidence of intent to discriminate on

racial grounds was presented. Similar grounds were
employed in refusing another argument, this time
involving the use of the referendum by the population
of a local government area. Article 34 of the
California Constitution was adopted by the population
in a referendum in 1950: it was placed before them as
the result of an initiative, termed the 'Public
Housing Project Law': its first section states that

> No low rent housing project shall hereafter be
> developed, constructed, or acquired in any
> manner by any state public body until a majority
> of the electors ... voting upon such issue,
> approve such project, by voting in favor thereof
> at an election ...

This was challenged on the grounds that by preventing the construction of low rent (mainly public)
housing projects in a municipality the article dis-
criminates against the poor as a class - thus denying them the equal protection of the laws - and
hence, because most blacks are poor, against blacks.

The Supreme Court refused to accept this argument, by a 6-3 majority, on two grounds: <u>James v
Valtierra</u> - 402 US 137, 28 L Ed 2d 678 (1971).
First, speaking for the majority, Mr. Justice Black
found no evidence of intent to discriminate by
race: 'the record here would not support any claim
that a law seemingly neutral on its face is in fact
aimed at a racial minority.' (p. 682)   Race,
according to the Court's majority in the 1970s, is
the only suspect classification (i.e. grouping of
the population) under the Fourteenth Amendment.
Mr   Justice Marshall disagreed:

> ... Article xxxiv on its face constitutes
> invidious discrimination ... It is ... an
> explicit classification on the basis of poverty
> - a suspect classification which demands
> exacting judicial scrutiny ...   It is far too
> late in the day to contend that the Fourteenth
> Amendment prohibits only racial discrimination;
> ... singling out the poor to bear a burden not
> placed on any other class of citizens tramples
> the values that the Fourteenth Amendment was
> designed to protect (pp. 684-5).

The second ground for dismissing the case concerned the use of the referendum.   On this, the
majority ruled that

Provisions for referendums demonstrate devotion
to democracy, not to bias, discrimination or
prejudice ... This procedure ensures that all
the people of a community will have a voice in
a decision which may lead to large expenditures
of local government funds ... This procedure for
democratic decision making does not violate the
constitutional command that no State shall deny
to any person "the equal protection of the laws"
(pp. 682-3).

Having formed a separate municipality, therefore,
and enacted a zoning plan designed to exclude 'undesirable' groups, a population could then use a
democratic device to ensure the continued exclusive-
ness of their community.   This ruling was later
extended in a case covering changes to zoning
schemes.   In 1971, the city of Eastlake, Ohio,
amended its municipal charter so that any proposal
to modify the land-use plan must first be approved
by 55 per cent of those voting in a referendum.
This, it was claimed by a property owner, was an unconstitutional delegation of legislative power to
the people which denied him due process.
        Six members of the Court disagreed.   Speaking
for that majority, Chief Justice Burger wrote that
'A referendum cannot ... be characterized as a
delegation of power.   Under our constitutional
assumptions, all power derives from the people'
(p. 137) and, echoing the James v Valtierra ruling,
'As a basic instrument of democratic government, the
referendum process does not, in itself, violate the
Due Process Clause of the Fourteenth Amendment when
applied to a rezoning ordinance' (p. 141).   In
dissent, Justices Powell, Stevens and Brennan
emphasised the difference between a referendum relating to a complete zoning scheme and one relating
to a rezoning of a particular piece of land only.
Previously, Courts had recognised only the validity
of the former - the cited case was Minneapolis -
Honeywell Regulator Co. v Nasady - 247 Minn 159
(1956) - but the Eastlake decision was extending it
to the latter.   The minority accepted the first,
but not the second: according to Mr Justice Stevens
(speaking also for Brennan):

I have no doubt about the validity of the
initiative or the referendum as an appropriate
method of deciding questions of community policy.
I think it is equally clear that the popular
vote is not an acceptable method of adjudicating

the rights of individual litigants (pp. 148-69).

Use of the referendum in such situations, according to this defeated view, denies the property owner the opportunity of a fair hearing for his case and, because it sets no clear standards, can be arbitrary: thus due process is denied, and the police power (see the Euclid decision) is being improperly used.

Standing to Challenge Exclusionary Zoning

The final case which emphasised the important role the 1970s majority on the Court allocated to local autonomy is a complex one relating to the exclusion- ary zoning practices of the town of Penfield, a suburb of Rochester, New York.   Four arguments were made against this practice:

1. developers and builders, represented by trade groups, complained that they were prevented from undertaking projects there - i.e. low- and moderate-income housing - and so had been deprived of potential profits;
2. low-income Rochester residents - of non- white ethnic stock - claimed they were pre- vented from living in an attractive suburban environment;
3. low-income residents of the City of Rochester claimed they were required to pay higher taxes because Penfield was not making a proportional contribution to the provision of low-income housing; and
4. several residents of Penfield claimed that they were unable to live in a racially integrated community.

The lower court had refused to accept these arguments, claiming that - under Article III, Section 2, of the Federal Constitution - the challengers had no standing to bring the case.   A 5-4 majority of the Supreme Court upheld this view - in Warth v Seldin 422 US 490, 45 L Ed 2d 343 (1975). Writing for the majority, Mr Justice Powell noted that

A federal court's jurisdiction ... can be invoked only when the plaintiff himself has suffered "some threatened or actual injury resulting from the putatively illegal action" ... (or) when the asserted harm is a

"generalized grievance" shared in substantially
equal measure by all or a large class of
citizens (pp. 354-5).

Neither the developers' and builders' associations,
the low-income residents of Rochester nor the
Penfield residents had standing under the interpretation.   The low-income residents did not 'allege
facts from which it reasonably could be inferred
that, absent the respondents' restrictive zoning
practices, there is a substantial probability that
they would have been able to purchase or lease in
Penfield' (p. 358) and the developers and builders
had 'failed to show the existence of any injury to
its members of sufficient immediacy and ripeness to
warrant judicial intervention' (p. 365).   Regarding
the argument concerning taxes in Rochester 'pleadings
must be something more than an ingenious academic
exercise in the conceivable ... We think the com-
plaint of the taxpayer-petitioners is little more
than such an exercise.' (p.360)   And the Penfield
residents did not show that they were unable to live
in a balanced community elsewhere.   Thus all four
groups were denied standing, as they could prove no
specific injury to themselves.
     The dissenting minority on the Court would have
let the case go to trial on its basic arguments.
Mr Justice Douglas argued that a 'clean, safe, and
well-heated home is not enough for some people.
Some want to live where the neighbors are congenial
and have social and political outlooks similar to
their own' (p. 366) and, on this basis, several of
the plaintiffs had standing, since the 'un-American
community model' of Penfield's zoning denied them
such an opportunity.   Mr Justice Brennan (joined
by Justices White and Marshall) went further,
arguing the paradox that success in enacting an
exclusionary zoning policy ensures that the policy
cannot be challenged (p. 369):

     The portrait which emerges from the allegations
     and affidavits is one of total, purposeful in-
     transigent exclusion of certain classes of
     people from the town ... the Court turns the
     very success of the allegedly unconstitutional
     scheme into a barrier to a lawsuit seeking its
     invalidation ... the Court tells the low-income
     minority and building company plaintiffs they
     will not be permitted to prove what they have
     alleged - that they could and would build and
     live in the town if changes were made in its

> zoning ordinance and its application - because
> they have not succeeded in breaching ... the
> very barriers which are the subject of the suit.

The majority had done this, he claimed, because of
an 'indefensible determination ... to close the
doors of the federal courts to claims of this kind'
(p. 372).

The Supreme Court's Rulings Reviewed

The final quotation from Mr Justice Brennan highlights the impact, if not the intent, of the
Supreme Court's rulings reviewed here.   During the
1970s in particular, when most of the cases concerning zoning and other land-use issues were considered, the Court presided over by Chief Justice
Burger has handed down majority decisions which
represent a narrow interpretation of the Fourteenth
Amendment - in particular of its equal protection
clause.   Only issues involving explicit intent to
discriminate on the basis of race have resulted in
rulings which override the autonomy of local governments, and in these it has been necessary for such
a discriminatory practice (as against implicit, via
a non-suspect classification, such as wealth) to be
proved.

   A minority has continued to argue for a wider
interpretation of equal protection and discrimination than with regard to race only, but this has
prevailed only very rarely.   Thus when in 1968
Mr Justice Marshall wrote that 'The ideals of
fairness, justice, and equality which prompted the
adoption of that amendment stand now as yet unful-
filled promises to new generations of men ... The
responsibility for making the amendment mean what
it says fall anew to each generation,' (Marshall,
1968, pp.1-2) he hoped that the next decade would
continue the fifteen or more years of judicial
activism in the Court presided over by Chief Justice
Warren.   He was to be disappointed: regarding the
issues discussed here, Lamb and Lustig (1979) have
concluded that 'the Burger Court's adherence to
judicial restraint and conservatism in exclusionary
zoning is often used as a mere guise to conceal its
general enmity to the substantive claims of indigent
minority plaintiffs.' (p. 226)   In seeking to avoid
offering relief to plaintiffs, the Burger Court -
largely fashioned by the appointments of Richard
Nixon - has upheld the rights of the residents of

independent local governments to control who lives
within the defined territory, so long as they do not
overtly and explicitly practice racial discrimination.   Thus, according to one legal commentator
(Sagar, 1978, p. 1425),

> the cases reflect the equation of the local
> zoning process with the joint exercise of the
> prerogatives of private ownership; the munici-
> pality is a club, which enjoys the mandatory
> and exclusive membership of its residents and
> landowners. And majority will - however insular,
> unjust, or irrational - prevail.

Another, referring to the same set of cases, concludes that 'the Supreme Court has gone about as
far as possible to eliminate the use of the United
States Constitution as a meaningful remedy for segregation (racial or low-income) (Pearlman, 1978,
p. 167).

These decisions favouring local autonomy, and
indicating that wealth is not a suspect classification under the Fourteenth Amendment even though its
use can have racial consequences, are consistent
with other Supreme Court judgements.   Regarding
school integration in metropolitan Detroit, for
example, the Court - in Milliken v Bradley 418 US
717, 41 L Ed 2d 1069 (1974) - overturned a lower
court decision that meaningful racial integration
could only be achieved if all 54 school districts
there were treated as a single unit.   In the
organisation of busing only the City of Detroit had
been acting illegally according to the Court
majority, and the autonomy of the other 53 (mainly
white suburban) districts should not be removed when
they had done nothing to deserve such penalty;   the
minority claim that school districts were created
by the State of Michigan, and this in itself involved the State in enabling discrimination, was dis-
missed (see Johnston, 1981c).   Similarly, in a case
alleging class discrimination against poor people
in the provision of education, because school districts differ widely in the available property tax
resources from which public education is financed,
a majority of the Court found - in San Antonio
School District v Rodriguez 411 US 1, 36 L Ed 2d 16
(1973) - that 'the class of disadvantaged "poor"
cannot be identified or defined in customary equal
protection terms' (p. 34) and (pp. 43-4)

It is not the province of this Court to create

> substantive constitutional rights in the name
> of guaranteeing equal protection of the laws
> ... the answer lies in assessing whether there
> is a right to education explicitly or implicitly
> guaranteed by the Constitution ... (there) is
> not.

Thus the system of separate (and unequal) school
districts was not ruled unconstitutional.

To some observers, all is not lost because the
United States Supreme Court has declined to rule on
many issues relating to land-use zoning and the
associated issue of local autonomy. These cases
have been left to the state courts, some of which,
notably in New Jersey (Moskowitz, 1978), have demonstrated considerable judicial activism in the consideration of zoning cases, and some aspects of
exclusionary zoning have been ruled unconstitutional.
One of the key cases - Southern Burlington County
NAACP v Township of Mt. Laurel (67 NJ 151, 336 A2d
713 (1975) - was initially welcomed as a major
defeat for such practices because it outlawed
exclusionary zoning and required municipalities to
take account of regional housing demands, but as
Sagar (1978, p. 1374) notes, 'even in the most
adventurous of these jurisdictions, New Jersey,
there has been a dance of substantive advance and
retreat which indicates a good deal of uncertainty
about the appropriate contours of the judicial
presence in land use planning.' (A full discussion
of the post-Mt. Laurel uncertainties in New Jersey
is given in Rose and Rothman, 1977.) Regarding
school finance, some state courts - in most cases
using their interpretations of the relevant state
constitution - have found inter-district differences
in available resources unconstitutional (despite the
San Antonio decision of the Supreme Court) and have
required equalisation plans to be introduced (see
Augenblick, 1979, and Lawyers' Committee for Civil
Rights under Law, 1980). Such decisions run
counter to the general judicial trend of not
threatening local autonomy, but they are exceptional
and affect only one aspect of the issue of separate
municipal incorporation and exclusionary zoning.

Conclusions

This essay has been concerned with two types of
institution in the United States - the local state
and the judiciary. Each is a creation of government

itself and as such has no independent existence.
Both could be removed, or substantially altered -
the first by legislative action at the state level
and the other by amendment of the Federal Constitution. Neither event is likely, however (although
several states now limit the further incorporation
of municipalities, especially close to existing ones:
Hallman, 1977), unless the institutions frequently
transgress the constraints to their powers - either
those explicitly stated or implicitly accepted.

This review of the interaction of the local
state and the judiciary in the United States over
the issue of land-use regulation has illustrated how
judicial interpretations of certain clauses of the
Federal Constitution have acted as constraints to
local activity. It has been shown that, except
where explicit racial discrimination and/or intent
to discriminate has been proved, all challenges to
local autonomy over the control of land use heard
by the Supreme Court have failed. Such failures
reflect the ideologies of a majority of the Supreme
Court justices, especially during the 1970s: for a
minority 'The framers of the Fourteenth Amendment
thought a solemn declaration would forever solve the
problems of social injustice' (Marshall, 1968, p. 9)
but such a belief could not be enshrined in new precedents. Until either the composition of the
Supreme Court changes or state and/or Federal
legislation directs otherwise, local state autonomy
for the manipulation of suburban spatial structure
is likely to remain.

Institutions such as the local state and the
judiciary are part of the superstructure of society.
Their existence, nature and functions can be studied
through theoretical analyses of the political
economy. Actions within such institutions - those
which influence spatial structures - result from
the decisions of individuals and groups, so that the
geographical study of institutions must include
detailed work on the managers and why and how they
make their decisions (Williams, 1978). In this way,
the operation of general processes can be uncovered
and the full impact of institutions identified.

Notes

1. Only late capitalist societies - as
exemplified by the United States - are considered
here. In socialist societies, the importance of
the state is even more crucial.

R.J. Johnston

2. The referencing system used here is that
commonly known as the Harvard system. The first
three entries (272 US 365) refer to the volume (272)
and first page number (365) of the Supreme Court
judgements; the next three (71 L Ed 303) refer to
the volume (71) and first page number (303) of the
annotated Lawyer's Edition. The last entry is the
year in which the Court delivered its judgement.
All references and quotations in this essay are to
the pages of the annotated edition.

References

Augenblick, J. (1979) School Finance Reform in the
    States: 1979, Education Finance Centre,
    Education Commission of the States, Denver,
    Colorado
Berger, R. (1977) Government by Judiciary, Harvard
    University Press, Cambridge, Massachusetts
Bigger, R. and Kitchin, J.D. (1952) Metropolitan Los
    Angeles: A Study in Integration. II How the
    Cities Grew, The Haynes Foundation, Los Angeles
Burnett, A.D. and Taylor, P.J. (eds.) (1981) Political
    Studies from Spatial Perspectives, Wiley,
    Chichester
Clark, G.L. (1981) 'Law, the State and the Spatial
    Integration of the United States', Environment
    and Planning A, 13, 1197-232
Cox, K.R. (1980) Location and Public Problems, Basil
    Blackwell, Oxford
Cox, K.R. and McCarthy, J.J. (1980) 'Neighborhood
    Activism in the American City: Behavioral
    Relationships and Evaluation', Urban Geography,
    1, 22-38
Cox, K.R. and McCarthy, J.J. (in press) 'Neighborhood
    Activism as a Politics of Turf: A Critical
    Analysis' in K.R. Cox and R.J. Johnston (eds.),
    Conflict, Politics and the Urban Scene, Longman,
    London
Danielson, M.N. (1976) The Politics of Exclusion,
    Columbia University Press, New York
Dear, M. (1981) 'The State: A Research Agenda',
    Environment and Planning A, 13, 1191-6
Dear, M. and Clark, G.L. (1978) 'The State and
    Geographic Process: A Critical Review',
    Environment and Planning A, 10, 173-83
Dear, M. and Clark, G.L. (1981) 'Dimensions of Local
    State Autonomy', Environment and Planning A, 13,
    1277-94

Dolbeare, K.N. (1969) 'Who Uses the State Trial
    Courts?' in J.R. Klonoski and R.L. Mendelsohn
    (eds.), The Politics of Local Justice, Little
    Brown, Boston
Dunleavy, P. (1980) Urban Political Analysis,
    Macmillan, London
Gordon, D.M. (1977) 'Class Struggle and the Stages
    of American Urban Development' in D.C. Perry
    and A.J. Watkins (eds.), The Rise of the
    Sunbelt Cities, Sage Publications, Beverly
    Hills, pp. 55-82
Greenberg, E.R. (1979) The Growth of Modern Govern-
    ment, Wiley, New York
Hallman, H.W. (1977) Small and Large Together, Sage
    Publications, Beverly Hills
Harvey, D. (1978) 'Labor, Capital and Class Struggle
    Around the Built Environment in Advanced
    Capitalist Societies' in K.R. Cox (ed.),
    Urbanization and Conflict in Market Societies,
    Maaroufa Press, Chicago, pp. 9-37
Hodder-Williams, R. (1980) The Politics of the U.S.
    Supreme Court, Allen and Unwin, London
Jackson, R.H. (1941) The Struggle for Judicial
    Supremacy, A.A. Knopf, New York
Jessop, B. (1977) 'Recent Theories of the Capitalist
    State', Cambridge Journal of Economics, 1, 353-
    73
Johnston, R.J. (1979) 'Governmental Influences in
    the Human Geography of "Developed" Countries',
    Geography, 64, 1-11
Johnston, R.J. (1980) City and Society, Penguin Books,
    Harmondsworth, Middlesex
Johnston, R.J. (1981a) 'The State and the Study of
    Social Geography' in P. Jackson and S. Smith
    (eds.), Current Research in Social Geography,
    Academic Press, London
Johnston, R.J. (1981b) 'The Political Element in
    Suburbia: A Key Influence on the Urban Geography
    of the United States', Geography, 66, 286-96
Johnston, R.J. (1981c) 'The Management and Autonomy
    of the Local State: The Role of the Judiciary
    in the United States', Environment and
    Planning A, 13, 1305-15
Johnston, R.J. (1982a) Geography and the State,
    Macmillan, London
Johnston, R.J. (1982b) The North American Urban
    System, St. Martin's Press, New York
Johnston, R.J. (in press) 'Voice as a Strategy in
    Locational Conflict: The Fourteenth Amendment
    and Residential Separation in the United States'
    in K.R. Cox and R.J. Johnston (eds.), Conflict,

R.J. Johnston

        Politics and the Urban Scene, Longman, London
Lamb, C.M. and Lustig, M.S. (1979) 'The Burger
        Court, Exclusionary Zoning and the Activist-
        Restraint Debate', University of Pittsburgh Law
        Review, 40, 169-226
Lawyers' Committee for Civil Rights under Law (1980)
        Update on State-Wide School Finance Cases, The
        Committee, Washington,DC
Marshall, T. (1968) 'The Continuing Challenge of the
        Fourteenth Amendment', Georgia Law Review, 3,
        1-10
Moskowitz, D.H. (1978) Exclusionary Zoning Litigation,
        Ballinger, Cambridge, Massachusetts
Muller, P.O. (1980) Contemporary Suburban America,
        Prentice-Hall, Englewood Cliffs, New Jersey
O'Connor, J. (1973) The Fiscal Crisis of the State,
        St. Martin's Press, New York
Pearlman, K.T. (1978) 'The Closing Door: The
        Supreme Court and Residential Segregation',
        American Institute of Planners Journal, 44,
        160-9
Perin, C.M. (1977) Everything in its Place,Princeton
        University Press, Princeton, New Jersey
Ranney, A. (1978) 'United States of America' in
        D. Butler and A. Ranney (eds.), Referendums,
        American Enterprise Institute for Public Policy
        Research, Washington,DC, pp. 47-86
Rose, J.G. and Rothman, R.E. (eds.)(1977) After Mt.
        Laurel: The New Suburban Zoning, Center for
        Urban Policy Research, Rutgers University, New
        Brunswick, New Jersey
Sagar, L.G. (1978) 'Insular Majorities Unabated',
        Harvard Law Review, 91, 1373-426
Saunders, P. (1979) Urban Politics, Penguin Books,
        Harmondsworth, Middlesex
Simon, J.F. (1973) In His Own Image, McKay, New York
Stetzer, D.F. (1975) Special Districts in Cook
        County, Research Paper 169, Department of
        Geography, University of Chicago, Chicago
Swindler, W.F. (1970) Court and Constitution in the
        Twentieth Century, Bobbs-Merrill, Indianapolis
Taylor, P.J. (1980) 'A Materialist Framework for
        Political Geography', Seminar Paper 37,
        Department of Geography, University of
        Newcastle-upon-Tyne, Newcastle-upon-Tyne
Taylor, P.J. (1981) 'Political Geography and the
        World-Economy' in A.D. Burnett and P.J. Taylor
        (eds.), Political Studies from Spatial
        Perspectives: Anglo-American Essays on
        Political Geography, Wiley, Chichester, pp.
        157-72

Teaford, J.C. (1979) City and Suburb, Johns Hopkins
    University Press, Baltimore
Tull, S.I. (1969) Zoned American, Grossman, New York
Walker, R.A. (1978) 'The Transformation of Urban
    Structure in the Nineteenth Century and the
    Beginnings of Suburbanization' in K.R. Cox
    (ed.), Urbanization and Conflict in Market
    Societies, Maaroufa Press, Chicago, pp. 165-212
Williams, N. (1950) 'Racial Zoning Again', American
    City, 65 (Nov.), 137
Williams, P. (1978) 'Urban Managerialism: A Concept
    of Relevance?', Area, 10, 236-40
Witt, E. (1979) Congressional Quarterly's Guide to
    the U.S. Supreme Court, Congressional Quarterly
    Inc., Washington,DC

Chapter 10

EDUCATION, INSTITUTIONS AND THE LOCAL STATE IN BRITAIN

Andrew Kirby

Education is the third greatest cause of human misery
in the world. The first, of course, is life.

Joseph Heller, Good as Gold

Introduction

Educational performance varies widely; between
social classes, between races and between localities
(Rutter and Madge, 1976). The reasons why these
variations occur are numerous, and it is probably
impossible to aim for a monocausal explanation. In
crude terms, however, we can identify two competing
models of attainment. The first may be characterised as a 'pure' theory, based upon intelligence,
whilst the second may be regarded as a 'secular'
model, taking into account educational inputs.
    Intelligence varies both socially and geogra-
phically, regardless of how it is defined or measured, but the complexity of studying genetic and
intergenerational effects suggests that a spatial
perspective may be a valuable starting point.
Robson, for example, has demonstrated intra-urban
IQ variations in Britain, as has Bailey for the
American city, albeit with a very small sample,
whilst Lynn has outlined the existence of what he
terms the 'social ecology of intelligence' (Bailey,
1980; Lynn, 1979; Robson, 1969). Lynn's work is an
attempt to account for variations in educational
attainment at the regional scale solely in terms of
population attributes. Such a perspective is rejected here; as I have argued elsewhere, it is not
possible to overlook variations in educational provision (Kirby, 1980; 1982), and this assertion is
developed below. Initially, the chapter will
examine spatial variations in educational provision
in Britain, and then variations in attainment,
evaluating the extent to which the two can be seen
to be related. Subsequently, various approaches
will be considered as satisfactory contexts within

which to examine these relationships, ranging from
the institutional to the structural.

A Geography of Education

In this section, the argument is based upon the
initial premise that educational attainment is a
function of educational provision, and as such side-
steps several issues - particularly the definition
of attainment; these issues are however rehearsed
in some detail in Kirby (1979a), and no further
justification will be made for linking attainment
and examination performance here.   Initially, we
may examine national variations of 'inputs' and
'outcomes'

The National Scale: Provision

It is generally accepted that, because education in
Britain is locally organised, there exist large
variations in provision between the units of
administration, the Local Education Authorities
(LEAs).   The extent of these variations is discussed in detail by Boaden (1971), Byrne, Williamson
and Fletcher (1975) and Pyle (1976), and it is clear
that the accident of birth within some LEAs can
entitle some children to the benefits of high
capital expenditure on all types of schools, and
high current spending on teachers and books: the
inverse of this unfortunately also applies.   The
extent to which current expenditure differs between
LEAs is indicated in Table 10.1, which reveals consistent variations between metropolitan and non-
metropolitan areas, and between inner and outer
London.

The National Scale: Attainment

Virtually all indicators suggest that educational
attainment also varies by region, and by LEA.
Harrop, for example, presents data for school-
leaving, examination success and further education,
and demonstrates that consistent patterns of under-
achievement exist (Harrop, 1976).   Charlton,
Rawstron and Rees concentrate solely upon the
number of Advanced Level examination students,
expressed as a proportion of each LEA's 18 year olds,
and show that the participation rates vary sharply
even between adjacent LEAs: the extremes are 9.5 per

Table 10.1: Local Authority Revenue Expenditure on Education, 1978-9

Local Authority	Expenditure[c]	Local Authority	Expenditure[c]
English Shire Counties and Constituent Districts (mean)			128.42
Avon	132.78	Isle of Wight	119.11
Bedfordshire	158.57	Kent	120.38
Berkshire	132.26	Lancashire	129.80
Buckinghamshire	143.23	Leicestershire	137.04
Cambridgeshire	128.78	Lincolnshire	121.59
Cheshire	141.32	Norfolk	112.40
Cleveland	154.65	Northamptonshire	126.85
Cornwall and Isles of Scilly	113.05	Northumberland	134.30
Cumbria	128.79	North Yorkshire	124.05
Derbyshire	123.50	Nottinghamshire	138.23
Devon	113.70	Oxfordshire	128.60
Dorset	116.31	Salop	124.17
Durham	130.93	Somerset	112.82
East Sussex	118.24	Staffordshire	140.27
Essex	122.28	Suffolk	115.52
Gloucestershire	129.94	Surrey	120.54
Hampshire	131.72	Warwickshire	124.50
Hereford and Worcester	125.83	West Sussex	104.68
Hertfordshire	152.07	Wiltshire	123.64
Humberside	136.40		
Welsh Counties and Constituent Districts (mean)			144.58
Clwyd	142.38	Mid Glamorgan	147.41
Dyfed	143.97	Powys	163.37
Gwent	141.64	South Glamorgan	139.39
Gwynedd	138.69	West Glamorgan	150.44
Metropolitan Counties and Districts[a] (mean)			142.60
Greater Manchester: Area Total	139.70	Tyne and Wear: Area Total	152.65
Merseyside: Area Total	146.01	West Midlands: Area Total	140.45
South Yorkshire: Area Total	145.09	West Yorkshire: Area Total	139.32
Greater London (mean)	156.44	ILEA[b]	195.84
Outer London Boroughs (mean)	134.72	Havering	127.48
Barking	146.85	Hillingdon	124.36
Barnet	134.79	Hounslow	138.72
Bexley	129.05	Kingston upon Thames	140.64
Brent	171.27	Merton	125.64
Bromley	129.48	Newham	156.27
Croydon	133.61	Redbridge	112.10
Ealing	143.94	Richmond upon Thames	121.57
Enfield	138.19	Sutton	119.84
Haringey	152.48	Waltham Forest	134.64
Harrow	135.60		
England and Wales (mean)	137.28		

Notes: a. Education is a responsibility of Metropolitan Districts; for comparative purposes, these have been aggregated by counties.
        b. Inner London Education Authority.
        c. Expenditure is calculated in £ per head of 1977 population.
Source: extracted from Chartered Institute of Public Finance and Accountancy (1979) Community Indicators, CIPFA, London.

cent and 33.9 per cent, with a national average of
approximately 18.3 per cent (Charlton et al., 1979;
Kirby, 1980).

The extent to which these two distributions (of
expenditure and results) can be easily related is
debatable. Current expenditure may be used in
many different ways, and may be selectively directed
within a single LEA. Similarly, different
authorities possess different legacies from past
expenditure: in 1973, for example, only 1 per cent
of primary pupils in Solihull were studying in
schools built before 1903; in Anglesey, 78 per cent
of primary pupils were in such old buildings. It
would be meaningless therefore to simply compare
maps of expenditure and results.

The only study which has attempted to systematically measure inputs (taking into account the ways
in which finances are distributed) and outcomes is
that of Byrne et al. The research, based upon 162
LEAs in 1970, was able to classify the authorities
into six types, as a result of applying cluster
analysis: each cluster reflects a particular
pattern of spending and aggregate achievement
(Byrne et al., 1975). The relative antiquity of
the results (which reflect now-defunct local
authorities and a variation in educational practice
that has also disappeared) militates against the
detailed discussion of the research: it is however
worth noting the authors' basic conclusions:
'spatially-defined variations in provision are
strongly related to spatially-defined variations in
measures of socially-significant educational
attainment' (Byrne et al., 1975, p. 155).

The Local Scale: Provision

Any gross analysis is open to problems of ecological
inference; as we have already noted, sums of money
can be spent in different ways: for the benefit of
all the pupils in an area, or perhaps to meet the
needs of a small minority. Indeed, studies of
individual education authorities reveal such distributional extremes. In the British case, Byrne
and Williamson have indicated the extent to which
different education authorities can place resources
in their schools in very different mixes and
amounts (1972). A suggestion that such variations
are commonplace in developed societies is also provided by Walker, in her study of Sydney (1979).
She provides details of significance tests undertaken

on various indicators of provision, and their variation between inner city and suburban wards; she
argues that the latter possess fewer disadvantaged
schools (i.e. old and poorly equipped), although
other aspects of provision (especially finance)
generally favour the inner areas.

An inner urban/suburban distinction in terms
of expenditure is a common one, which reflects the
poor state of many old, central schools, and the
heavy demands frequently placed upon them.   In
Britain, Government policy has at times been to encourage LEAs to bus immigrant children away from
such schools, because the latter were unable to
cope with their particular educational needs (Edwards
and Batley, 1978, p. 30).   More recently, specific
patterns of discriminatory aid have been focused
upon poor schools in areas of high need, via the
Educational Priority Area scheme;   the extent to
which this additional expenditure has been in any
context successful remains to be seen however
(Williamson and Byrne, 1979).

The Local Scale: Attainment

As we reduce the spatial scale in our examination
of attainment, we must of course pay greater regard
to the variations in social class composition,
which become more marked between small areas, and
which can distort performance statistics.   None
the less, even when this social segregation is taken
into account, attainment can be shown to possess a
spatial dimension.

In his study of literacy in Inner London, for
example, Panton notes that the city's eight year
olds scored on average slightly below the national
mean; the test is evaluated to yield a national
average of 100, with a 10 point divergence represen-
ting a year's difference in reading ability.   The
Inner London average is 94.4, but as Table 10.2
indicates, the variance about this figure is quite
large (Panton, 1980, p. 30).   In general, children
from non-manual homes have higher scores than do
children from other backgrounds, but the picture is
consistently distorted by spatial effects.   Thus
the gap between the mean scores for 'non-manual'
children and 'unskilled' children in area N is 20.8
points - a gulf of about two years.   However, the
range within the non-manual averages from area to
area is larger: 21.1 points.   Indeed, location can,
it seems, compensate for parental background: a

Table 10.2: Social Class of Parent and Spatial Variations in Literacy, London, 1968

Area	Non-manual mean	Skilled mean	Semi-skilled and unskilled mean	All children mean
E. Lewisham	105.2	101.4	103.0	102.6
P. Lewisham	107.3	97.7	93.2	102.4
R. Greenwich	105.9	99.0	92.5	94.9
S. Greenwich	106.6	101.7	93.4	101.2
D. Islington	100.3	98.5	98.3	95.5
G. Greenwich	112.2	106.4	*	109.1
H. Greenwich	105.1	98.3	90.2	97.0
I. Greenwich	*	97.3	94.7	95.8
L. Southwark	*	106.4	93.4	98.2
N. Lewisham	108.4	103.9	87.6	101.0
Q. Greenwich	105.8	97.2	89.9	94.6
A. Southwark	93.0	93.6	94.0	93.3
B. Greenwich	104.1	90.8	92.3	93.9
C. Lewisham	103.6	96.0	93.3	98.2
F. Lewisham	94.2	94.6	93.8	93.4
J. Hackney	*	93.9	92.0	93.7
M. Southwark	101.8	95.0	91.9	93.2
U. Lewisham	105.0	94.1	89.6	95.1
K. Tower Hamlets	91.1	90.4	87.3	88.5
O. Tower Hamlets	93.9	94.5	88.4	90.2
J. Hackney	100.7	91.4	85.8	88.8
All Children	104.6	96.8	91.3	95.8

Note: All scores higher than the column average are underlined; where lack of data makes a mean unreliable, this is recorded *. Column headings refer to the social class background of children.
Source: extracted from Panton, R.E. (1980) 'Literacy in London', Geography, 65, 27-34

comparison of areas K and E, for instance, reveals that in the latter children from potentially the poorest social groups are on average a year older in reading age than children with professional and managerial parents in area K.

The material presented here is consistent with the premise outlined in the Introduction, namely that there exist spatial variations in provision, and also spatial variations in attainment. Ultimately, to causally relate the two distributions, we need to show that there is some mechanism whereby performance can be 'bought' by an injection of funds into the educational system, and - equally important - some explanation as to why funding should vary. The only realistic assumption is that this process occurs - if at all - via individual schools: consequently it is with this proposition that the next section begins.

Organisation

The School

It is relatively easy to show that different schools achieve different levels of aggregate performance. Figure 10.1, for example, is a comparison of nine comprehensive schools in Newcastle-upon-Tyne, with respect to four measures of attainment. This display of variation is not unusual, although relatively difficult to account for. Superficially, it may appear that social mix alone is responsible for these variations, insofar as the highest levels of attainment are associated with schools drawing on higher-income, owner-occupied neighbourhoods. However, it is possible to take these factors into consideration and still find differences, as the sophisticated strategy employed by Rutter et al. indicates (Rutter, Maughan, Mortimore and Ouston, 1979).

Figure 10.2 indicates two things. First, it contrasts mean examination scores for children in each of twelve comprehensives: all are within the Inner London Education Authority. (The scores are produced by aggregating the results of all the fifth-year pupils in each school, and allocating one point for an O-level pass, and half a point for each CSE pass.) Second, the diagram also disaggregates the results in terms of the initial abilities of the children on entry to the school: this is measured in terms of the Verbal Reasoning Quotient (VRQ), and

Andrew Kirby

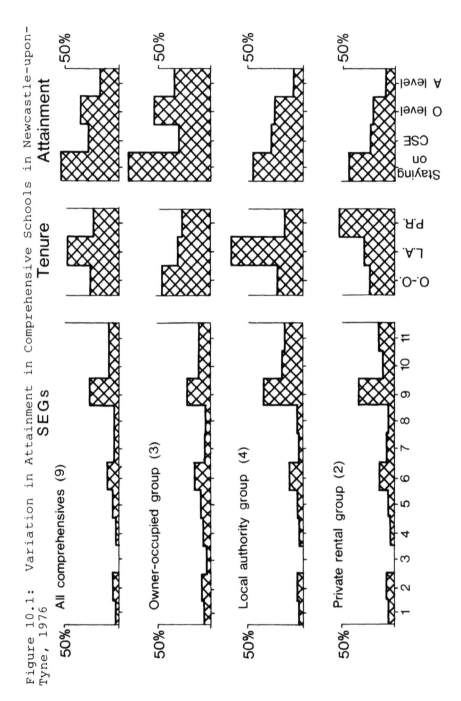

Figure 10.1: Variation in Attainment in Comprehensive Schools in Newcastle-upon-Tyne, 1976

Figure 10.2: Aggregate Attainment in 12 ILEA Comprehensive Schools, Disaggregated by Scores of Children at Entry on VRQ Tests

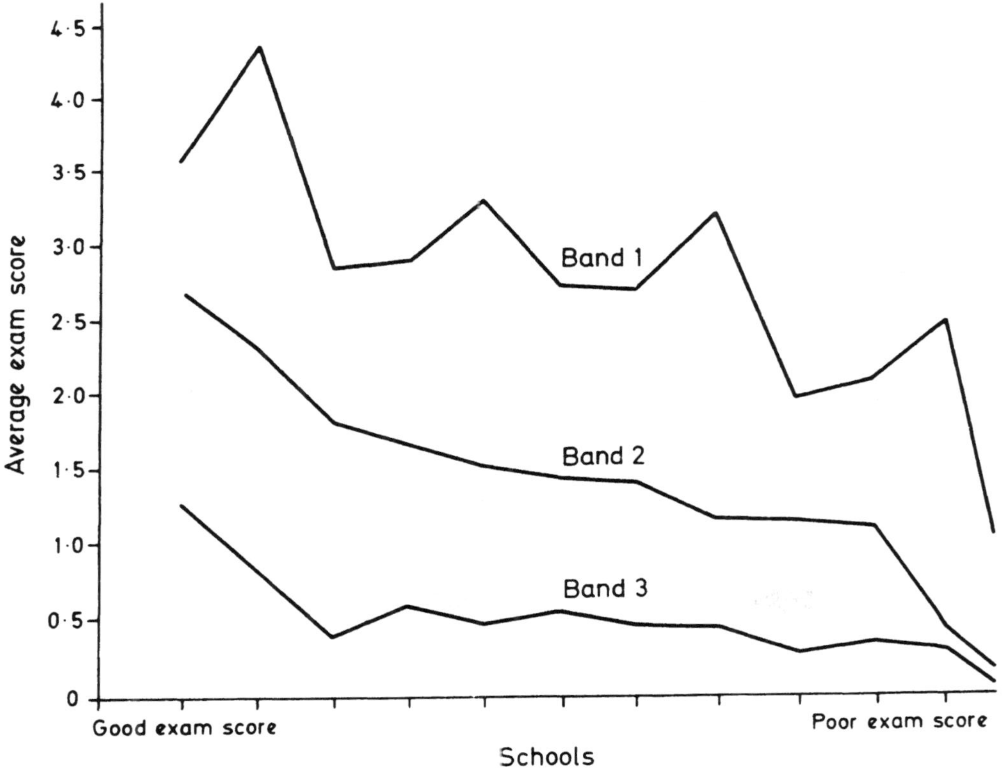

Andrew Kirby

performance has been simplified to one of three
'bands' - Band 1 being the highest.   As we might
expect, children with greater abilities on entry (at
eleven years) perform most successfully at sixteen;
in each school, Band 1 children maintain their lead.
What is of greater significance is that attendance
at an individual school can boost the probabilities
of academic success.   Thus in some comprehensives,
Band 1 children perform little better than do Band 3
children in others: in other words, attainment levels
in some schools are on average far higher than in
others, even when factors such as the initial
abilities of the pupils are taken into account.
     The material discussed in Figures 10.1 and 10.2
indicates two things therefore.   The first is that
different schools offer different probabilities of
success to their pupils, and the second is that this
process can be shown to occur even when the abilities
of the children are taken into account.   In consequence, it is not surprising that the organisation
of local educational provision becomes both administratively and politically important, in that the
decision to direct certain children to certain
schools may lead to a particular level of attainment
that will not be achieved elsewhere (such organisational questions also have a long political history
in the USA, as Cox demonstrates: 1979, p. 276).
Furthermore, this decision remains important because
a 'successful' school - in the sense used here -
cannot be simply created.   Rutter et al. show that
physical criteria of organisation, such as the age
of the buildings or whether the school is scattered
over several sites, actually count for very little;
instead the success of the school will depend upon
its long term development as a 'social organisation'
(Rutter et al., 1979, pp. 106-44).

The Spatial Organisation of School Systems

Essentially, there exist three ways in which
children from particular neighbourhoods are assigned
to particular schools; the decision may rest with
the parents, or alternatively some form of spatial
districting may be employed, involving catchment
areas or a feeder system.   Primary education is
generally simpler, insofar as primary schools are
usually local schools, although the drawing of catchment boundaries may also be problematic.
Secondary schools are necessarily larger and therefore some form of spatial organisation is required:
Table 10.3 indicates the proportions in which the

298

Table 10.3: Comprehensive School Allocation Systems
in England and Wales, 1977, by Local Education
Authority

Allocation system	proportion of LEAs employing system
Catchment	51%
(involving some parental choice)	(19%)
Feeder system	19%
(involving some parental choice)	(11%)
Parental choice alone	27%

Source: Dore, C. and Flowerdew, R. (1981)
'Allocation Procedures and the Social Composition of
Comprehensive Schools', Manchester Geographer, 2,
No. 1, Table 1

three alternative systems are employed in England
and Wales.

In each of the three cases, it is possible to
argue that inequalities may arise. Feeder systems,
whereby children from several primary schools enter
a particular secondary school, are discussed by the
author elsewhere, especially with respect to the way
in which they give rise to concentrations of children
from some neighbourhoods in certain schools; in
simple terms, the feeder system can be used to per-
petuate the social distinctions that existed between
the academically strong grammar schools and the
academically weak secondary modern schools (Kirby,
1979a,b).

The issue of catchment systems has been examined
in Derby by Dore and Flowerdew (1981). Within the
city, four sectors exist, and each operates a particular catchment system. In three of the sectors,
little attempt has been made to take account of the
social geography of Derby, and schools tend to
serve only one neighbourhood. It is only in Sector
C that a detailed attempt has been made to produce
some form of social mixing. As Figure 10.3 indicates,
Homelands, Derby and Littleover schools each draw
upon a local, suburban catchment. The inner city
is however an 'option zone', and pupils from this
area have a choice of any of the three schools (the
fourth school, Sinfin, is oversubscribed). The net
result is that each school serves a full range of
neighbourhoods, although it is worth re-emphasising
that this has necessitated thoughtful 'spatial
engineering'.

Clearly, in LEAs where either feeder or catchment systems operate, the opportunities for individual children to attend the strongest schools may be
limited. This may suggest that an entirely free
system, in which parents have complete control of
the choice of school, is desirable. Dore and
Flowerdew disagree: they argue that 'the schools
with better reputations [are] effectively creaming
many of the middle-class pupils from the remainder'
(1981). This is not surprising. As we shall see
below, education is within some strata of society
(particularly the middle- and high-income groups and
religious and ethnic minorities) a sensitive issue;
consequently it is these groups which will take the
greatest trouble to search out the best schools.
Where parental choice is unavailable, those that are
able buy their way into the catchment areas of these
schools. As Tiebout has argued, in such cases local
resources can simply be captured by the most affluent

Figure 10.3: Comprehensive School Catchment Areas in Derby

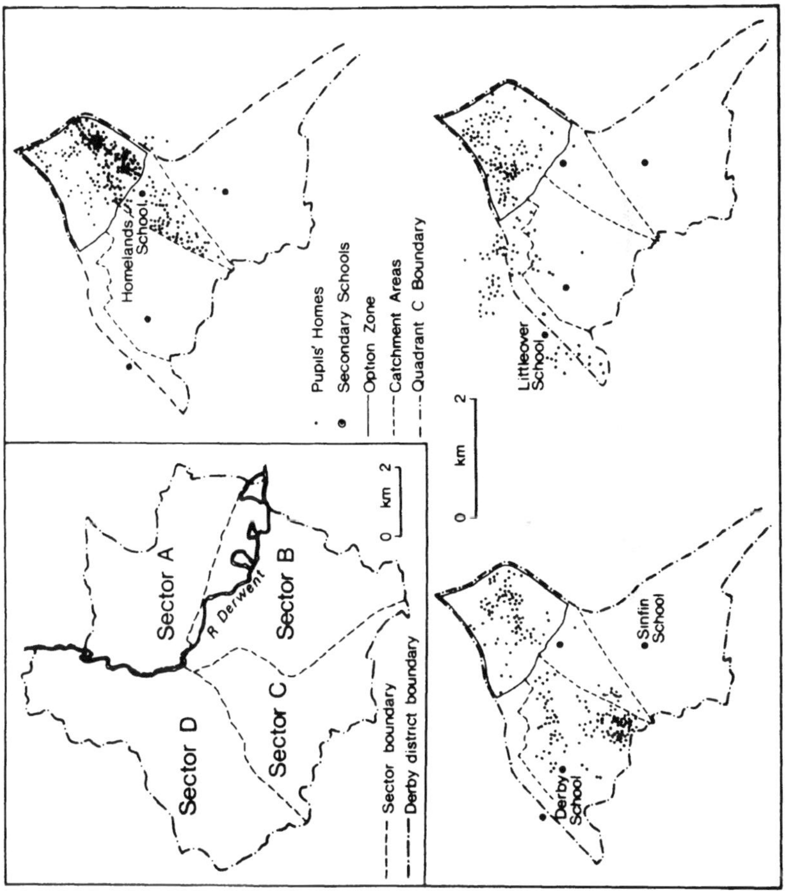

Andrew Kirby

(Forrest, 1981; Tiebout, 1956).
    In all three systems therefore, it is possible
to argue that the spatial organisation of local
education provision restricts individual
opportunities; more particularly, it is possible to
show examples whereby social inequalities are being
reinforced by the school system - simply, a class
division exists between schools, with predictable
implications for levels of attainment.   This leaves
us with the problem of explaining the existence of
such types of organisation, which clearly oppose the
egalitarian intent of comprehensive (non-selective)
education.

The Local Education Authority

The study of particular policies invites us to
examine the decision-making process at work within
the education system.   As studies such as that by
Kogan and van der Eyken illustrate, the administrative chain is a complicated one, although for our
purposes it can be reduced to three main groups of
actors: the government body responsible for education, the Department of Education and Science (DES);
the elected members, who collectively constitute the
local Education Committee, and the Chief Education
Officer (CEO) (Kogan and van der Eyken, 1973, pp.
25-52).
    Much of the literature within the field of
administration is keen to stress that the DES has
little effective control over local educational
practice (Pyle, 1976).   This general argument is a
seductive, albeit - it will be argued below - now
a dated view.   Within the context of this discussion, the specific issue of school districting
has however traditionally been a local concern, and
consequently attention will be focused there for the
present.
    In his discussion of educational administra-
tion, Brooksbank outlines two goals that are normally
to be met when allocating pupils to schools.   The
first is that parents' wishes should be identified:
he continues, 'when parental preferences have been
received, the authority is faced with the task of
allocating children fairly to the schools available.
Schools which are undersubscribed present no problem.
Any child living within the area who requires a place
there can be admitted.   If a school is oversubscribed ... the principle generally adopted is that
those to whom the school is most accessible are given

302

a place there.   Geographical proximity is the first
criterion' (Brooksbank, 1980, p. 190).

In this case, then, we can see that social goals
do not figure as one of the intended outcomes of
pupil allocations.   This is unsurprising, and
accords with the growing stock of studies which may
be described as 'managerialist' (Pahl, 1975).
Increasingly, it is becoming clear that many professionals and professional organisations venerate
abstract concepts such as 'efficiency' and 'planning'
far more than the achievement of any social aims.
Brooksbank's book, although written on behalf of the
Society of Education Officers, which represents the
profession of educational administration, contains no
statements as to the purpose of the education system,
merely broad but meaningless goals, such as 'the
constant search for more appropriate, better and
surer ways of educating all the children within our
schools' (Brooksbank, 1980,p. 39).

This is not to argue that inequitable systems of
organisation arise solely because CEOs are unaware of
the social implications involved.   In some instances,
the elected members of the Education Committee may
direct policy, and a system change dictated by the
chairman of such a Committee is discussed in Kirby
(1979a, pp. 90-4).   In the case in question, policy
was shifted to the creation of a feeder system which
supposedly met the contradictory aims of creating
community schools and reducing social polarisation;
in fact it simply increased the latter.

This notwithstanding, the study of professional
administrators is a useful first step in the analysis
of any policy field.   In this instance, a study of
the Society of Education Officers makes it clear
that CEOs subscribe to no particular ethos (comprehensive education, core curricula, sixth-form
colleges) and have only a limited interest in these
non-technical issues: 'educational planning ought to
involve ... things which either should be attempted
or which those in power are determined to attempt
(the two are often the same thing)' (Brooksbank,
1980, p. 44).   Moreover, the involvement of elected
members in decision-making is unwelcome: the CEO and
the Education Committee chairman 'are regularly
stimulated, indeed often nettled, by a flow of advice
and criticism from practitioners and public'
(Brooksbank, 1980, p. 45).

Where such a focus however fails is in its concentration upon superficial appearances (Kirby,
1979a). A narrow consideration of the professional
ethos need not question the existence of inequalities

between schools (no discussion is offered either by
Brooksbank), nor the initial evidence that some
LEAs choose to spend much more upon education than
do others.   Equally, it is enough to admit the
occasional 'interference' of elected members in the
policymaking process, without attempting to recognise the source of their views and allegiances.
In simple terms, an emphasis upon the administration
of education depoliticises what is essentially a
political activity.   As a result, an alternative
approach to a simplistic investigation of the
managers of the system is required.

Education and the Local State

In his study of the limitations of 'liberal'
approaches to (urban) geography, Gray argues (1976,
p. 39):

> the liberal geographer would presumably examine
> the socio-spatial distribution of educational
> opportunity; take an institutional approach
> and investigate the allocation of educational
> resources by central and local government ...
> or, study the selection of children for
> different types of school ... However, this
> approach ... is subject to all the criticism
> made of non-radical geography. It is essential-
> ly descriptive and, by its very nature, deals
> only with the surface outcomes of deeper
> structural processes.

After his demolition of the liberal view
(which has of course been pursued in the first parts
of this chapter), Gray outlines the necessary contexts within which to comprehend the role of education within the capitalist state.   Essentially,
the state is involved in education as part of the
process of labour reproduction, a role that has increased in importance as the needs of industry and
commerce have become more demanding in terms of
skills and qualifications.   In addition however,
the state education system also acts in an ideological manner (through the inculcation of social values)
and as a means of perpetuating income and status
division, via the distribution of different levels
of educational attainment (Gray, 1976, pp. 40-2).
The value of this approach is that it gives a
richer understanding of the geographical patterns
(and other phenomena) that we encounter.   If we

follow Byrne et al., it becomes relatively easy to
account for the variations that exist between LEAs
in terms of expenditure, and more importantly,
attainment; they argue that 'education will reflect
what key groups in society regard as important
educational aims, and these in turn will reflect
what such groups regard as important social and
economic needs. Historically, such needs have
always been defined in labour market terms.'
(Byrne et al., 1975, p. 45) We can reinterpret
results (such as those outlined by Harrop) therefore
as a reflection of the educational needs of a region
being translated into varying levels of aggregate
attainment, via the educational system; regions with
declining industries, a history of manual labour and
low female activity rates will not demand high levels
of skill, particularly from women, and there will be
no political pressure, from any direction, for high
levels of educational investment. A simple comparison of attainment levels in the South East and
Northern regions of England is sketched in Table
10.4

The Local State

As we have already argued, education is to be understood as an aspect of local provision; in consequence, we need to emphasise not simply the role of
the national state, but also that of the 'local
state'. The meaning of the latter term is open to
some confusion, but it is generally used to imply a
study of sub-national government (including local
authorities, health authorities and so on) from a
materialist perspective. An excellent review is
provided by Dear (1981), and Boddy (1980) has also
outlined the issues of current concern with respect
to an understanding of the local state. These are:
the development of critical theory; the importance
of local struggles; and the pattern of centre-local
relations. These will be discussed in turn in the
context of education.

Local Struggles. It has already been noted
above that the relationship between CEOs and the
'consumer' is an ambivalent one, at least from the
former's standpoint. From other perspectives, the
importance of local, community pressures upon educational organisation can be large. It has been
argued here that education is a sensitive issue,
particularly for high-income groups and religious
and ethnic minorities. In each instance, there

exist numerous examples in which parents opt for 'specialised' education - either in denominational schools or private institutions (Forrest, 1981; as we might expect, the national distribution of wholly-private schools mirrors the spatial distribution of high socio-economic groups throughout the country as a whole - Harrop, 1976).

There also exist examples in which the term 'political struggle' can be more appropriately used. Saunders, for example, documents the opposition within Croydon, South London, to DES plans to intro-duce comprehensive education in 1965 (1980, pp. 262-6). Following the announcement of comprehensive re-organisation, parental opposition mounted and, to their delight, this caused the Education Committee to rescind its plans: 'the leader of the council explained this sudden reversal in terms of deference to public opinion. "It is our duty to do as you would wish" he told the Purley parents, and that is exactly what we have done' (1980, p. 264). As Saunders points out, this example must be qualified in two respects. The first is that community opposition was congruent with the basic political outlook of the Education Committee, and consequently was successful because it voiced a generally-held view: this of course underlines the fact that political struggles are usually unequal contests. As both Cockburn and O'Malley show in their studies of working-class neighbourhoods in London, many struggles in the educational field tend to focus on peripheral questions - such as playspace and pre-school playgroups - simply because community action finds it difficult to penetrate the professional aura which cloaks the central issues of education, such as the quality of schools. In consequence, local energies become directed into schemes which improve neighbourhood facilities, but which do little to erode basic inequalities in provision (Cockburn, 1977; O'Malley, 1977).

The second issue shifts the focus of attention away again from the LEA. In the example of Croydon, it should be remembered that the source of the unpopular comprehensivisation plans was the central administration, and not the Local Education Authority.

Centre-local Relations. The accepted wisdom suggests that educational policy has owed little to central control, and points out that comprehensive schooling, for example, emerged within some LEAs, in opposition to DES wishes:

Table 10.4: Comparisons of Attainment, Northern and South East Regions, 1971-2; Percentage of All School Leavers

ATTAINMENT		NORTH	SOUTH EAST	ENGLAND AND WALES
No	O-levels	61.1	51.7	56.3
1-2	O-levels	11.9	13.1	12.1
3-4	O-levels	7.3	8.8	8.0
5+	O-levels	19.8	26.4	23.6

Source: Harrop, K.J. (1976) 'An Educational Profile of the Northern Region in Comparison with Other Regions', Working Papers No. 25-6, North-east Area Study, University of Durham

there is considerable scope for autonomy on the part of LEAs, in deciding on education policy. Not only in matters of style - e.g. type of secondary system provided, the content of the curriculum and the age of transfer from primary to secondary schooling, but also in terms of the amount of resources used e.g. teaching staff, age and standards of buildings, equipment and facilities (Pyle, 1976, p. 111).

The issue of central-local relations is thus a confused one; as we have seen immediately above, Saunders discusses the comprehensive issue in Croydon from a perspective which emphasises the ultimate financial sanction of the central state (1980, p. 265):

when the members of the Education Committee and their bureaucratic advisers were summoned to Whitehall in April 1967, and informed that central government finance for virtually all future planned capital expenditure on schools in the town was to be withheld until such time as a suitable comprehensive plan was submitted, it became clear that the issue was for all intents and purposes resolved. Not even Croydon's articulate and influential middle class could hope to take on the central government and win.

Clearly, finance is the key to the relationship between the national and local state, and it can be argued that the former will intercede to limit local spending whenever the latter threatens to promote a widespread fiscal crisis: 'the significance of the degree of control retained by the central state is manifest in the fiscal crises of cities' (Dear, 1981, p. 198). In the British context, the Rate Support Grant dictates the overall level of expenditure open to local agencies, although more specific tactics may be employed. A significant move is the proposal to introduce a core curriculum to apply throughout England and Wales (curricular content, it will be remembered, was regarded by Pyle as one of the Local Education Authority's major freedoms). Plans to limit the bulk of teaching to a core of subjects such as mathematics, English, science and a language have obvious implications for the proliferation of courses in schools in fields such as sociology, environmental science, domestic economics and the arts, and thus also implications for levels of expenditure on teachers and equipment (in parentheses, it is to be noted that geography is not scheduled to appear in this core of subjects - DES, 1979).

This specific proposal is thus one aspect of an apparent increase in the state's involvement in local affairs, which has ranged from attempts to dismantle the Inner London Education Authority (a unique organisation which involves a group of LEAs acting in concert) to major changes in local housing strategies. As Boddy has noted, it remains to consider these trends - if trends they are - from a theoretical standpoint.

Critical Theory. The development of theory with respect to the local state is still nascent; consequently, it is not surprising that our abilities to apply this understanding to the field of education are severely limited. None the less, it is possible to generate some general conclusions. These are as follows:

1. Education is one aspect of reproduction, managed at the local scale, and financed - in part at least - to fulfil local needs and to maintain local levels of attainment. The growth of technical education for school-leavers is to be seen in this light (Kirby, 1979b, pp. 24-7).
2. Educational organisation at LEA scale has traditionally seen some autonomy, but relations between the central state and the local state appear to be changing, with the former asserting greater financial control.
3. Political activists may influence local decisions, particularly when they are middle class in nature and sophisticated in organisation. The failure to implement egalitarian allocation systems within LEAs must be seen as consistent with the bulk of these demands, namely to maintain exclusivity in certain schools.
4. The local state also has relations with local economic activities, and these frequently will be of an instrumentalist nature. Many private educational establishments are relatively wealthy institutions with commensurate political influence. Local governments are likely to welcome private schools on the grounds that they

contribute to the area's rates, and absolve
some of the LEA's responsibilities for
education (the 1980 Education (2) Bill
provides central funds to finance private
education).   Both Dunleavy and Saunders
document the growth of private schooling
and, more particularly, the attempts made
by private schools to obtain preferential
planning treatment when moving from central
areas to high-status greenfield sites
(Dunleavy, 1980, pp. 82-6; Saunders, 1980,
pp. 249-58; see also Marsden for a historical dimension: 1980, p. 24).

This list is not exhaustive; it does not for
example include potential areas of interest, such as
the political protests likely to develop in rural
areas as a result of the erosion of accessibility to
schools, and in other LEAs hit by falling school
rolls, with all that this implies for the closure of
schools.   The problems facing ethnic minorities
have only been touched upon (Reading Council of
Community Relations, 1980) and no attention has been
paid to the relations between LEAs over busing
arrangements for pupils, which may cause administrative acrimony.   These issues notwithstanding, it
seems clear that an emphasis upon the local state,
and the internal and external relations that it
generates, provides a rich basis for the development
of theory and the consequent interpretation of
educational issues.

Conclusions

It remains to adduce some general themes from this
chapter.   The general aim has been to document and
account for some of the gross variations that exist
in terms of educational practice and aggregate
attainment at all levels and at various spatial
scales.   The basic premise has been that these two
patterns are causally related, and it is argued that
the evidence presented here is consistent with that
interpretation.
A more difficult problem is faced when attempts
are made to explain these geographical patterns.
A narrow emphasis upon the actors involved - the DES,
the LEAs, the CEOs - is helpful to a degree, but
must remain of limited value.   In consequence, a
wider perspective has been sought, and found, in an
emphasis upon the operation of the local state.

Such a view emphasises the commitments of the LEA to the local economy, its relations with the state, and its responses to local political demands. At present, our understanding of the local state is limited; the potential is however great, particularly with respect to the geography of education.

Acknowledgements

Acknowledgements are due to Robin Flowerdew for permission to use material in Figure 10.3, and Michael Rutter for material used in Figure 10.2.

Some of the material discussed in this chapter was originally presented to a workshop held at the School for Advanced Urban Studies, University of Bristol, in December 1980. I would like to thank the participants, Michael Dear and Martin Boddy for their comments on other but related papers, and Robin Flowerdew and Sophie Bowlby for suggestions concerning this chapter. All shortcomings remain my responsibility.

References

Bailey, W.H. (1980) 'The Spatial Pattern of Intelligence in a Small Town', Journal of Geography, 79, 174-80

Boaden, N. (1971) Urban Policy Making, Cambridge University Press, Cambridge

Boddy, M. (1980) 'The Local State: Theory and Practice', unpublished paper, School for Advanced Urban Studies, University of Bristol

Brooksbank, K. (ed.) (1980) Educational Administration, Councils and Education Press, London

Byrne, D. and Williamson, W. (1972) 'Some Intra-regional Variations in Educational Provision and their Bearing upon Educational Attainment - the Case of the North-East', Sociology, 6, 71-87

Byrne, D., Williamson, W. and Fletcher, B. (1975) The Poverty of Education, Martin Robertson, Oxford

Charlton, W.A., Rawstron, E.M. and Rees, F.E.H. (1979) 'Regional Disparities at A-level', Geography, 64, 26-36

Chartered Institute of Public Finance and Accountancy (1979) Community Indicators, CIPFA, London

Andrew Kirby

Cockburn, C. (1977) The Local State, Pluto Press, London

Cox, K.R. (1979) Location and Public Problems, Blackwell, Oxford

Dear, M. (1981) 'A Theory of the Local State' in Burnett, A.D. and Taylor, P.J. (eds.), Political Studies from Spatial Perspectives, Wiley, Chichester, pp. 183-200

Department of Education and Science (1979) A Framework for the Curriculum, Her Majesty's Stationery Office, London

Dore, C. and Flowerdew, R., (1981) 'Allocation Procedures and the Social Composition of Comprehensive Schools', Manchester Geographer, 2, No. 1, 47-55

Dunleavy, P. (1980) Urban Political Analysis, Macmillan, London

Edwards, J. and Batley, R. (1978) The Politics of Positive Discrimination, Tavistock Press, London

Forrest, J. (1981) 'Spatial Implications of the Journey to School: A Review and Case-study', Pacific Viewpoint, May

Gray, F. (1976) 'Radical Geography and the Study of Education', Antipode, 8, No. 1, 38-44

Harrop, K.J. (1976) 'An Educational Profile of the Northern Region in Comparison with Other Regions', Working Papers No. 25-6, North-east Area Study, University of Durham

Kirby, A.M. (1979a) Education, Health and Housing, Saxon House, Farnborough, Hampshire

Kirby, A.M. (1979b) 'Towards an Understanding of the Local State', Geographical Papers No. 70, Department of Geography, University of Reading

Kirby, A.M. (1980) 'A Comment on the Social Ecology of Intelligence in the British Isles', British Journal of Social and Clinical Psychology, 19, 333-6

Kirby, A.M. (1982) 'Public Issue or Private Achievement? A Further Comment on the Issues of the Social Ecology of Intelligence and Educational Attainment', British Journal of Social Psychology, 21, 63-7

Kogan, M. and van der Eyken, W. (1973) County Hall: The Role of the Chief Education Officer, Penguin Books, Harmondsworth, Middlesex

Lynn, R. (1979) 'The Social Ecology of Intelligence in the British Isles', British Journal of Social and Clinical Psychology, 18, 1-12

Marsden, W.E. (1980) 'Travelling to School: Aspects of Nineteenth-century Catchment Areas',

312

Geography,65, 19-26

O'Malley, J. (1977) The Politics of Community
Action: A Decade of Struggle in Notting Hill,
Spokesman, Nottingham

Pahl, R.E. (1975) Whose City? 2nd edn., Penguin
Books, Harmondsworth, Middlesex

Panton, R.E. (1980) 'Literacy in London', Geography,
65, 27-34

Pyle, D. (1976) 'Aspects of Resource Allocation by
Local Education Authorities', Social and
Economic Administration, 10, 106-21

Reading Council of Community Relations (1980)
'Structural Inequality in School Allocation',
Multiracial Education, 9, 26-8

Robson, B.T. (1969) Urban Analysis, Cambridge
University Press, Cambridge

Rutter, M. and Madge, N. (1976) Cycles of Disadvantage, Heinemann, London

Rutter, M., Maughan, B., Mortimore, P. and Ouston, J.
(1979) Fifteen Thousand Hours, Open Books,
London

Saunders, P. (1980) Urban Politics, Penguin Books,
Harmondsworth, Middlesex

Tiebout, C. (1956) 'A Pure Theory of Local Expenditure', Journal of Political Economy, 64, 416-24

Walker, S., (1979) 'Educational Services in Sydney:
Some Spatial Variations', Australian Geographical Studies, 17, 175-92

Williamson, W. and Byrne, D. (1979) 'Educational
Disadvantage in an Urban Setting' in Herbert,
D.T. and Smith, D.M. (eds.), Social Problems and
the City, Oxford University Press, Oxford,
pp. 186-200

SUBJECT INDEX

AUTHOR INDEX

Form, W.H. 13
Forrest, J. 302, 306
Forsyth, F. 70
Found, W.C. 76
Freeman, A.M., III
143, 153-4
Friedmann, J. 169, 188
Friends of the Earth
123

Gaile, G.L. 169, 175,
188
Galbraith, J.K. 30
Gale, S. 35, 51, 53,
54, 56, 57, 59-61,
66
Gaskin, M. 91
Gauldie, E. 236
Geisler, C.C. 87
Georgescu-Roegen, N.
53, 60
Gerstl, J.E. 16
Gibson, K. 232
Gilbert, A.G. 174
Gilg, A.W. 76, 87
Gladwin, T.N. 142, 146
Gleave, D. 211
Glickman, N.J. 170
Glyn, A. 22
Gober, P. 212
Godsden, P.H. 236, 238
Goodman, D.E. 174
Goodwin, C.D. 154
Gordon, D.M. 264
Gordon, P. 176, 185-6,
191
Gordon, W.C. 31
Gough, I. 22, 169
Gough, T.J. 238-9
Government Research
Corporation 170
Grafton, D. 89
Graham, G. 146
Grant, M. 25
Gray, F. 11, 13, 16,
26, 230, 304
Green, L.P. 24
Greenberg, E.R. 255
Greenwood, M.J. 212
Gregory, D. 232
Greytak, D. 169

Grigg, D. 57
Guldman, J.M. 152

Haber, S.E. 217
Habermas, J. 53
Haddon, R.F. 14-15
Hall, P. 88, 95
Hall, R.K. 213
Hallman, H.W. 261, 283
Hansen, K.A. 210
Harloe, M. 17-19, 26
Harrison, B. 179
Harrison, M. 245
Harrop, K.J. 290, 305-7
Hart, S.L. 144
Harvey, D. 23, 59, 193,
233-4, 262
Hatch, J.C.S. 16
Healy, R.G. 87, 162
Henderson, J. 249
Her Majesty's Stationery
Office 94
Herbert, D. 248
Hinderink, J. 169
Hirschman, A.O. 173, 188-9,
192
Hodder-Williams, R. 259,
269
Hofstadter, R. 17
Holen, A.S. 219
Holland, S. 59, 169, 179,
198
Holloway, J. 22-3, 194
House of Commons, Select
Committee on Energy 135
House of Lords, Select
Committee on the
European Communities 103,
123
Housing Monitoring Team
242, 245-7
Hudson, J.R. 233
Hudson, R. 169
Huggett, F. 78
Hughes, J.W. 179
Hutton, S.D. 16

Ingram, H.M. 105, 119
IUCN, 104
Isard, P. 57
Isard, W. 57